EXPLANATORIUM OF THE EARTH

DK

EXPLANATORIUM OF THE EARTH

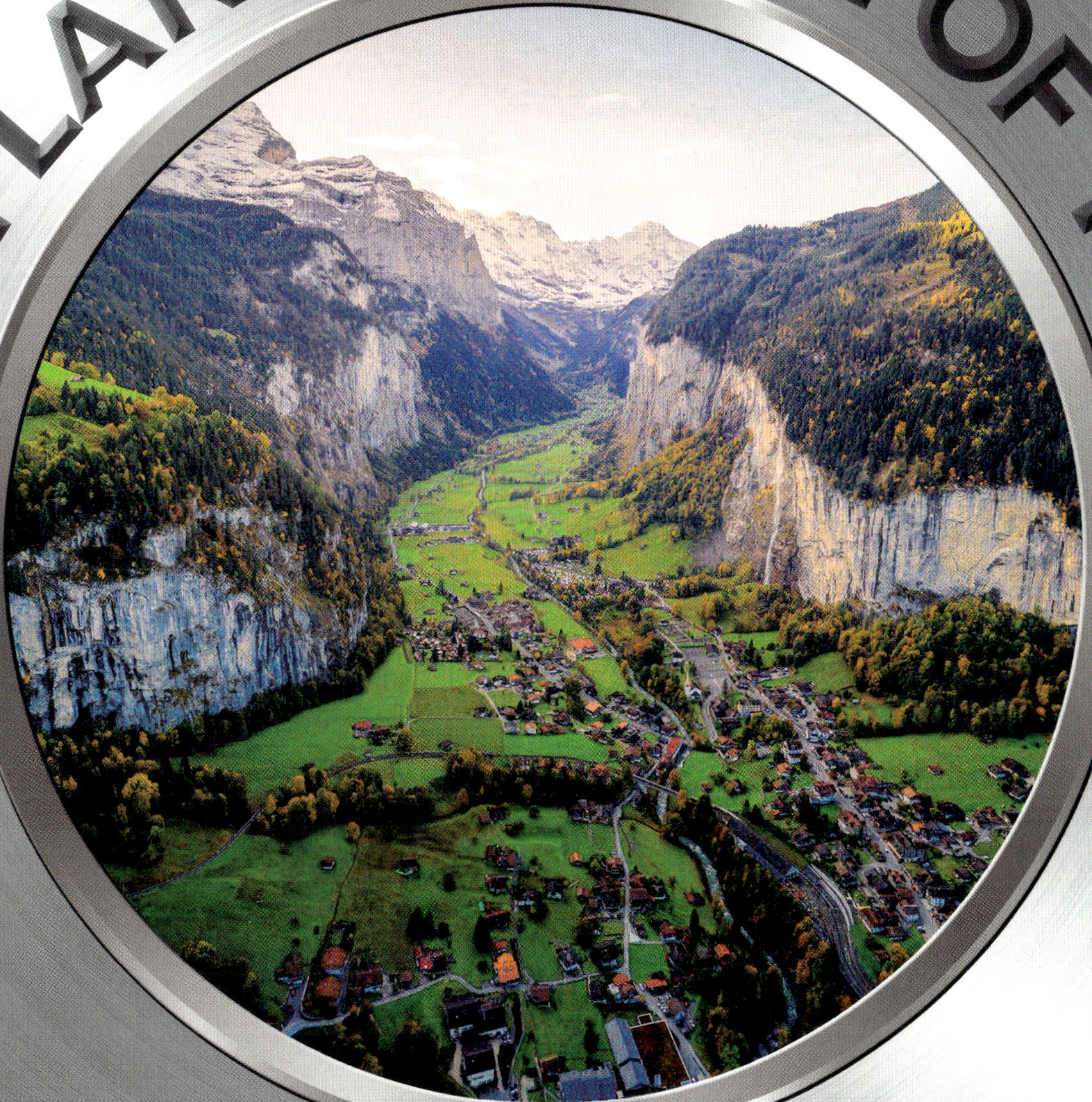

CONTENTS

DK LONDON
Senior editor Ben Morgan
Senior art editor Smiljka Surla
Editors Joseph Barnes, Shaila Brown, Jolyon Goddard, Orso Publishing
Designers Tory Gordon-Harris, Anna Pond, Rhys Thomas
Illustrators Peter Bull, Sofian Moumene, Simon Mumford, Simon Tegg
Photographers Ruth Jenkinson, Gary Ombler
Creative retouching Steve Crozier
Picture researcher Laura Barwick
Managing editor Rachel Fox
Managing art editor Owen Peyton Jones
Production editor Rob Dunn
Production controller Meskerem Berhane
Jackets design development manager Sophia MTT
Senior jackets coordinator Priyanka Sharma Saddi
Jacket designer Stephanie Cheng Hui Tan
Senior DTP Designer Harish Aggarwal
Publisher Andrew Macintyre
Art director Karen Self
Publishing director Jonathan Metcalf

Authors
Joseph Barnes, Abigail Beall,
Professor Joseph Holden, Wendy Horobin,
Dr Peter Innes, Dr Kate Ravilious,
Giles Sparrow, Isabel Thomas,
Dr Rebecca Williams

Consultants
Professor Joseph Holden, Dr Peter Innes,
Professor Stephen Marshak,
Cally Oldershaw, Dr Rebecca Williams

First published in Great Britain in 2024 by
Dorling Kindersley Limited
DK, One Embassy Gardens, 8 Viaduct Gardens,
London, SW11 7BW

The authorised representative in the EEA is
Dorling Kindersley Verlag GmbH. Arnulfstr. 124,
80636 Munich, Germany

Copyright © 2024 Dorling Kindersley Limited
A Penguin Random House Company
10 9 8 7 6 5 4 3 2 1
001–334051–May/2024

A CIP catalogue record for this book
is available from the British Library.
ISBN: 978-0-2416-0166-2

Printed and bound in China

www.dk.com

MIX
Paper | Supporting
responsible forestry
FSC™ C018179

This book was made with Forest
Stewardship Council™ certified
paper – one small step in DK's
commitment to a sustainable future.
Learn more at
www.dk.com/uk/information/sustainability

EARTH IN SPACE

INSIDE THE PLANET

VOLCANOES AND EARTHQUAKES

CHANGING LANDSCAPES

ROCKS AND MINERALS

THE ATMOSPHERE

THE BIOSPHERE

REFERENCE

Earth belongs to a family of eight planets orbiting the Sun, our **nearest star**. The Sun, planets, moons, and millions of smaller bodies like **comets** and **asteroids** make up our **solar system**. All the solar system's planets formed about 4.5 billion years ago in a series of **collisions**. These violent events shaped our world and its orbit, giving us our Moon, 24-hour days, 365-day years, seasons, and tides.

HOW **EARTH** FORMED

Planet Earth formed around 4.5 billion years ago from a swirling cloud of space debris that encircled the newborn Sun. Over millions of years, the tiny particles of matter in this cloud collided with each other and gradually clumped together, growing into whole planets.

▶ SPACE ROCKS

Meteorites are the oldest rocks known to science and contain the raw materials that formed our planet. The Imilac meteorite crashed into the Atacama Desert in Chile about 700 years ago. It consists of the metals iron and nickel – the same metals found in Earth's core – and is studded with yellow-green crystals of olivine – the main mineral found in Earth's mantle (the layer between the planet's crust and core). Scientists think the Imilac meteorite was once part of a planet or asteroid that was smashed by a collision in the solar system's early years.

BIRTH OF THE SOLAR SYSTEM

The Solar System began to form when an interstellar gas and dust cloud (a nebula) began to shrink due to gravity. The Sun formed about 4.6 billion years ago and the planets developed from the cloud of debris that surrounded it.

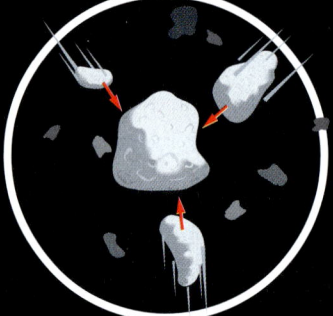

1 SWIRLING CLOUD
A giant cloud of gas and dust collapses. The force of gravity pulls the debris into a dense, spinning disc. Some of the lighter gases are flung to the cold outer regions of the cloud.

2 THE SUN
The dense core of the disc becomes so hot, it triggers a nuclear reaction and a star – our Sun – is born. The leftover debris forms a spinning disc around the Sun.

3 PLANETESIMALS
Particles of dust and rock within the spinning disc collide and clump together, forming massive objects called planetesimals. One of these will become Earth.

4 NEW PLANET
The planetesimals repeatedly crash into each other. These collisions heat and melt their interiors, and gradually, an irregularly shaped Earth forms.

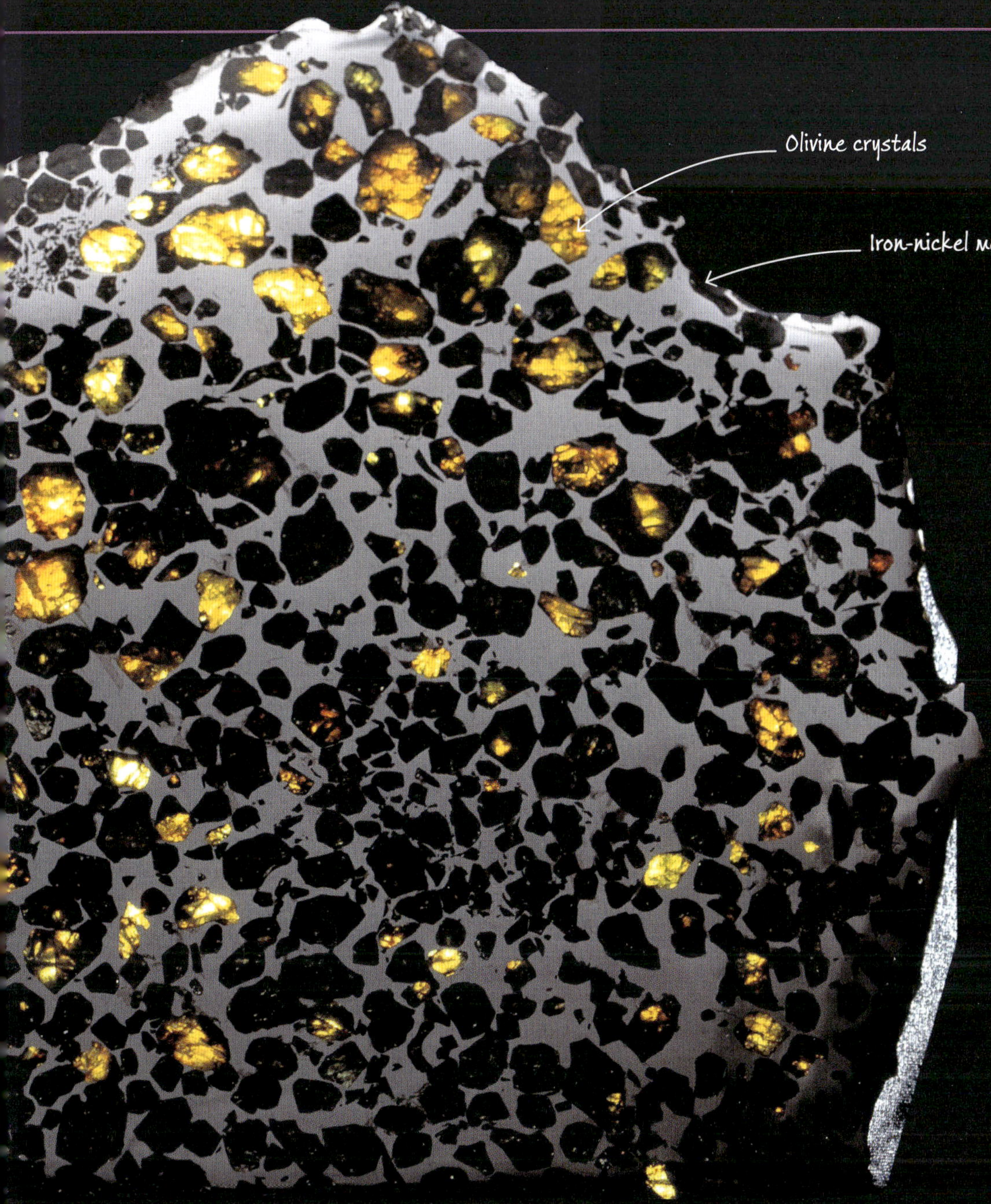

Olivine crystals

Iron-nickel metal

When small meteors collide with Earth's atmosphere, they burn up and create shooting stars.

The Moon was much closer to Earth than it is today.

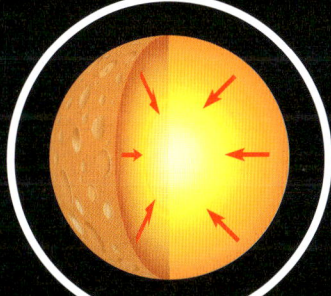

5 EARLY EARTH
The inward pull of gravity reshapes Earth into a sphere. Heavier materials like iron and nickel sink to form the core. Lighter materials like rock minerals form the mantle.

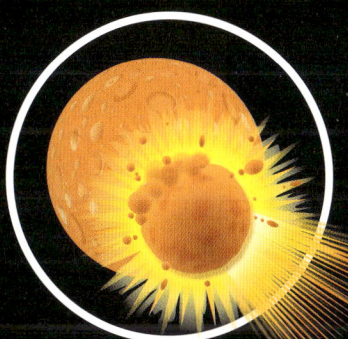

6 COLLISION COURSE
Late in Earth's formation, a collision with a small planet creates a cloud of debris around Earth. The collision also gives Earth's axis its tilt.

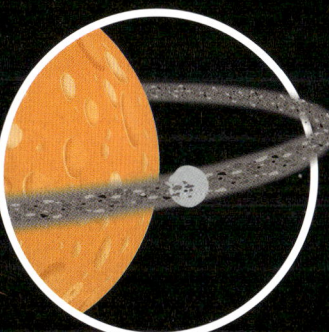

7 NEW MOON
The debris from the collision forms a ring around early Earth. The material – a mixture of rock and metal – clumps together to form the Moon.

8 CHANGING EARTH
Outpouring of volcanic gases – composed mainly of carbon dioxide and water – creates early Earth's atmosphere. As Earth cools and moisture condenses, it rains to form the oceans.

IMPACT CRATER

Violent collisions with space rocks are less common today than in Earth's early years, but the threat still exists. Tenoumer Crater formed as recently as 20,000 years ago when a huge meteorite smashed into the Sahara desert, leaving a scar 1.9 km (1.2 miles) wide. Large impact craters are rare on our planet because erosion and weathering soon wear them away. The best preserved craters are found in deserts.

HOW THE SOLAR SYSTEM WORKS

The solar system is the area of space dominated by our local star, the Sun. Earth is the third of eight major planets that orbit (travel around) the Sun, trapped by the pull of its gravity. The solar system is also home to hundreds of moons, more than a million asteroids, and countless icy objects called comets.

▼ SOLAR MODEL
This simple model made from a ball and marbles shows the order of the planets in the solar system. The four inner planets are balls of rock and metal. The outer planets are gas giants – giant spinning globes of hydrogen and helium.

VENUS
Venus is about the same size as Earth but is smothered by a dense atmosphere of toxic gases and clouds of sulphuric acid. Carbon dioxide traps heat to make it the hottest planet in the solar system.

MARS
Mars is the outermost of the rocky planets and a little less than half the diameter of Earth. It is a bitterly cold, desert world but shows signs of being warmer and wetter in the distant past.

THE SUN
The Sun is a vast ball of hot gas containing 99.8 per cent of the solar system's mass. The tremendous force of gravity from this great mass keeps the rest of the objects in the solar system trapped in orbit around it.

MERCURY
The smallest planet, Mercury has a huge metal core and a crater-covered surface.

EARTH
Earth is just the right distance from the Sun for water to exist as a liquid on the surface, making life possible. Earth is also the only planet with a crust broken into moving tectonic plates. Their motion creates volcanoes, mountains, and Earth's varied landscapes.

DISTANCE FROM SUN
Distances in space can be hard to grasp intuitively. If the Sun was the size of a basketball at one end of a basketball court, Earth would be a grain of sand at the other end. The Earth–Sun distance is known as one astronomical unit.

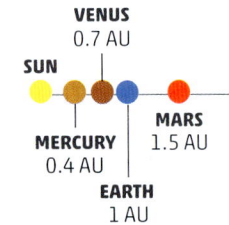

VENUS
0.7 AU

SUN

MERCURY
0.4 AU

MARS
1.5 AU

EARTH
1 AU

JUPITER
5.2 AU

SATURN
9.5 AU

ORBITS

Objects in the solar system travel around the Sun along paths called orbits. Orbits are not perfectly circular. Instead they have shapes known as ellipses, which vary from highly elongated to nearly circular. The major planets all orbit the Sun in the same plane – the plane formed by the ring of debris that surrounded the newborn Sun 4.6 billion years ago. Smaller objects, including the dwarf planet Pluto, have more elliptical orbits that are tipped relative to the plane of the planets.

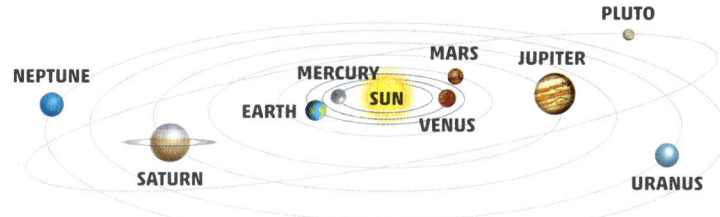

COMETS

Comets swoop in and out of the inner solar system on highly elliptical orbits. When a comet is near the Sun, its icy surface evaporates and forms a tail.

ASTEROID ITOKAWA

ASTEROIDS

These giant rock vary from a few metres to hundreds of kilometres wide. They have irregular shapes and are mostly found in the asteroid belt, a zone between the orbits of Mars and Jupiter.

JUPITER

Jupiter is the largest planet and has more than 90 moons. It is a fast-spinning globe of hydrogen and helium, wrapped in colourful bands of windswept cloud. These include the Great Red Spot, a storm large enough to swallow Earth.

SATURN

Saturn is a gas giant with more than 140 moons and a spectacular system of rings. The rings consist of trillions of chunks of debris from the catastrophic destruction of an icy moon or comet millions of years ago.

SOLAR POWER

The Sun is powered by nuclear fusion reactions in its core. These generate light energy, which sustains life on Earth. It takes thousands of years for light to travel from the Sun's core to its surface but only eight minutes to reach Earth.

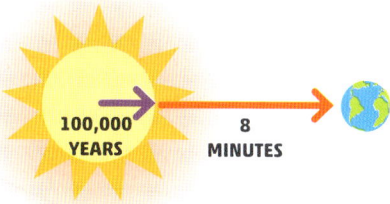

100,000 YEARS   **8 MINUTES**

URANUS

Uranus is an ice giant with a hydrogen- and helium-rich atmosphere surrounding layers of slushy chemicals. Due to an interplanetary collision long ago, it spins like a rolling ball, with a horizontal axis of rotation.

NEPTUNE

Neptune is another ice giant and the outermost major planet. Despite receiving little energy from the Sun, it has surprisingly active, stormy weather and the highest wind speeds in the solar system – up to 2,000 kph (1,200 mph).

URANUS
19 AU

NEPTUNE
30 AU

DAY AND NIGHT

Earth's daily rotation creates the natural cycle of light and shade we know as day and night. A whole day is 24 hours, but the number of daylight hours varies depending on where you live and the time of year. These variations happen because Earth spins round on a tilted axis.

▶ SPINNING GLOBE

Earth rotates eastwards on an imaginary line called an axis. As the planet turns, different parts of the surface face the Sun and receive daylight, while other parts are in shadow, causing night. Because Earth's axis is tilted, the number of sunlight hours varies from place to place, ranging from as little as zero hours a day to 24 hours a day.

At the poles, the Sun sets and rises only once each year. There are six months of continuous daylight followed by six months of twilight or darkness.

Earth's axis is tilted 23.5° from upright in relation to the planet's orbit around the Sun. As a result, day and night are a different length in most places.

Milan in Italy is halfway between the equator and the North Pole. Here, the number of daylight hours varies from about 9 hours in midwinter to nearly 16 hours in midsummer.

Points on Earth's surface move east as the planet rotates. As a result, the Sun rises in the east and sets in the west.

At the Equator, people get 12 hours of daylight and 12 hours of darkness every day of the year. Sunrise is always around 6 am and sunset is around 6 pm.

DAYS AND YEARS

A day lasts 24 hours from noon to noon, but it takes Earth only 23 hours, 56 minutes, and 4 seconds to rotate once. The times are different because Earth moves a little along its orbital path round the Sun over the course of a day. As a result, it has to rotate slightly more than one turn to face the Sun directly again. Over a year, these extra bits of rotation add up to more than a full turn. Earth actually rotates 366.24 times a year, even though there are only 365 days.

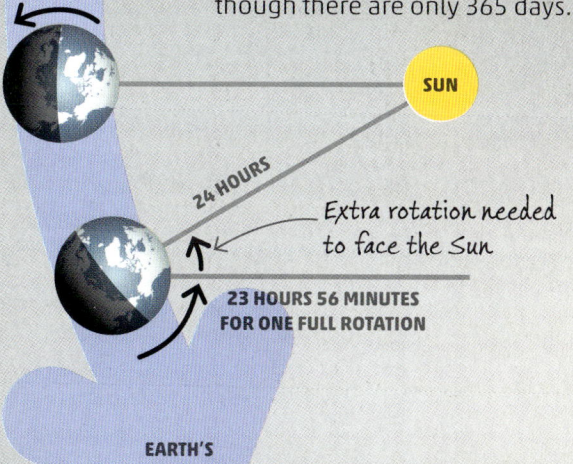

SUN

24 HOURS

Extra rotation needed to face the Sun

23 HOURS 56 MINUTES FOR ONE FULL ROTATION

EARTH'S ORBIT

EARTH'S ROTATION

In the past, people watched the Sun and stars crossing the sky and naturally concluded that they travel around Earth. One of the first people to realize that Earth was rotating was the Indian mathematician Aryabhata in the 6th century. He also accurately calculated the length of one rotation as 23 hours, 56 minutes, and 4 seconds.

CHANGING DAYS

Powerful earthquakes can change the way Earth spins. In 2004, a huge earthquake in the Indian Ocean rocked the whole planet, shifting the North Pole by about 2.5 cm (1 in) and reducing the length of a day by 2.7 microseconds.

EPICENTRE OF 2004 EARTHQUAKE

SPEEDY EQUATOR

Because Earth is spherical, different places on its surface move at different speeds as the planet rotates. The poles are stationary, but the Equator whizzes round at about 1,600 kph (1,000 mph). Rockets are often launched from near the Equator to give them an extra boost so they can reach orbit more easily.

LAUNCH OF ARIANE ROCKET IN FRENCH GUIANA, NEAR THE EQUATOR

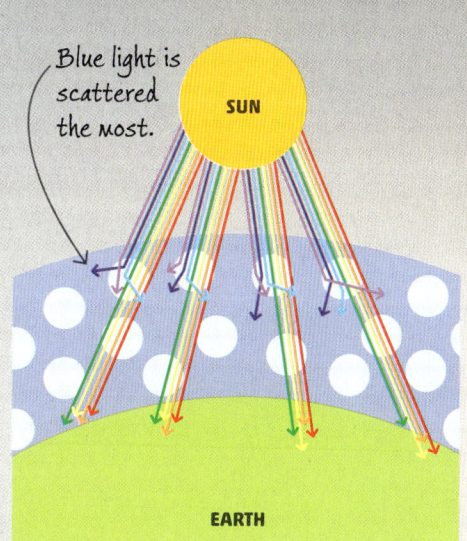

Blue light is scattered the most.

SUN

EARTH

WHY IS THE SKY BLUE?

At night the sky is black, revealing the vastness of space, but during the day it turns bright blue. This happens because the brilliant light from the Sun is scattered by air molecules in Earth's atmosphere. White light is a mix of all the colours of the rainbow, but blue light is scattered more easily than other colours. The scattered blue light makes the whole sky look blue.

▶ **AROUND THE SUN**

Earth orbits the Sun once a year. As it goes, each day it spins around an imaginary line called an axis, which runs from pole to pole. But the axis tilts at an angle of 23.5° and, as a result, the northern and southern hemispheres lean towards the Sun at different times of the year, causing the cycle of seasons.

NORTHERN SPRING

Around 20 March, Earth's axis is at right angles to the Sun. Days and nights are about the same length everywhere in the world. It is spring in the northern hemisphere and autumn in the southern hemisphere.

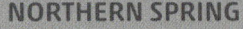

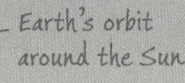

Earth's orbit around the Sun

NORTHERN SUMMER

The longest day of the year in the northern hemisphere is around 21 June (summer solstice). Around this time, the northern hemisphere tilts directly towards the Sun, causing long, warm days and short nights. But in the southern hemisphere, because the Earth tilts away from the Sun, this is the shortest day of the year (winter solstice).

SUN

HOW
SEASONS
WORK

Most of the world experiences seasons. As the months pass, the weather gradually changes from cold to warm or from dry to wet and then back again. These seasons are not caused by Earth getting closer to the Sun or further away. They are the result of our planet's tilt.

NORTHERN AUTUMN

Around 23 September, instead of tilting towards the Sun, Earth's axis is at right angles to the Sun, and days and nights are about the same length everywhere. It is now autumn in the northern hemisphere and spring in the southern hemisphere.

The equator is an imaginary line between the northern and southern hemispheres.

MIDNIGHT SUN
Places at the equator get roughly equal hours of day and night all year round, but near the poles, the seasonal variation between day and night is extreme. At midsummer in the Arctic, there is no night as the Sun stays above the horizon 24 hours a day. Meanwhile, in Antarctica, the Sun doesn't rise and it stays dark for weeks on end.

NORTHERN WINTER
The darkest day of the year in the northern hemisphere is around 21 December and is called the winter solstice. This is when the northern hemisphere tilts directly away from the Sun, causing the short days and cold weather of winter. The southern hemisphere tilts towards the Sun and so experiences summer.

By December, the South Pole tilts towards the Sun, bringing summer in the southern hemisphere.

FOUR SEASONS
In spring, as days start to get longer, leaves begin to appear and flowers start to bloom. The long days and the warmth of summer coincide with plants growing quickly. In autumn, the days start to shorten and many leaves turn red or orange, before falling. In winter, when it is coldest and the days are shortest, most plants are bare. Then, it's spring once again.

SPRING

SUMMER

AUTUMN

WINTER

TROPICAL SEASONS
Countries near the equator don't have the four seasons. However, most tropical countries have a dry and wet season. In tropical parts of the northern hemisphere, the wet season takes place at the same time as the northern summer, and the dry season happens in the northern winter. The opposite happens in the southern hemisphere.

DRY SEASON

WET SEASON

SEASONAL SURVIVORS
Baobab trees have adapted to severe dry seasons in the tropics. In the wet season, their massive trunks store thousands of litres of water for the dry season.

HOW ECLIPSES WORK

Every so often, the Sun, Earth, and the Moon line up directly in space. When the Moon passes between Earth and the Sun, it casts a shadow on Earth and we see a solar eclipse. When Earth casts a shadow on the Moon, we see a lunar eclipse.

▶ TOTAL SOLAR ECLIPSE

The most spectacular kind of eclipse is a total solar eclipse. For a few minutes, day turns almost to night and stars appear as the Moon crosses the Sun and blocks its light. You need to be in the centre of the Moon's shadow to see a total solar eclipse. In other locations, people see a partial eclipse.

By a strange coincidence, the Sun is 400 times wider than the Moon and 400 times further away. As a result, the two appear the same size in our sky and fit almost perfectly during a total solar eclipse.

DIAMOND RING

All total solar eclipses begin as partial eclipses, as the Moon starts to make its way across the Sun. Just before the Sun is completely blocked, the last rays of sunlight pass through valleys on the Moon, creating a bright spot known as a diamond ring. This spectacular effect only lasts for a few seconds.

PARTIAL AND ANNULAR ECLIPSES

Partial solar eclipses happen when only some of the Sun's disc is blocked. If the Moon isn't perfectly aligned with Earth and the Sun, the Sun appears as a crescent. An annular eclipse happens when the Moon lines up perfectly but is slightly further from Earth than normal and doesn't completely cover the Sun.

PARTIAL

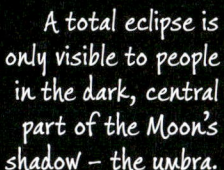

A total eclipse is only visible to people in the dark, central part of the Moon's shadow – the umbra.

In the outer, paler part of the Moon's shadow (the penumbra) people see a partial eclipse.

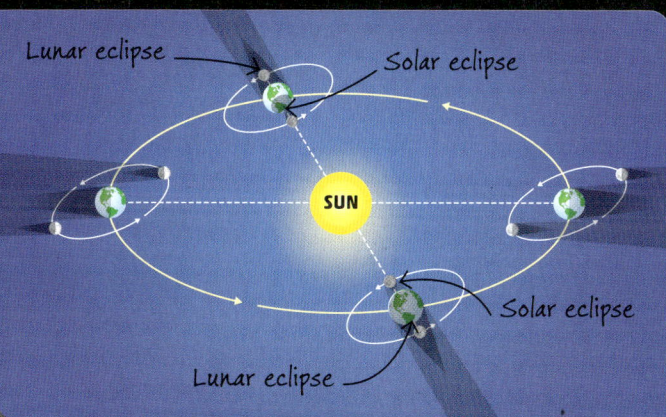

Lunar eclipse

Solar eclipse

SUN

Solar eclipse

Lunar eclipse

TILTED ORBIT
If the Moon orbited Earth in the same plane as Earth orbits the Sun, we'd have eclipses every month. However, the Moon's orbit is tilted a few degrees compared to Earth's, so it is usually too high or too low. Where it does cross the Sun-Earth lines, a total solar or lunar eclipse happens.

TOTALITY
During a total solar eclipse, the sky darkens and birds stop singing. The Sun's brilliant disc disappears entirely but its outer atmosphere – the corona – becomes visible as a glowing halo around the Moon. It's dangerous to look at the Sun directly, so never watch a total solar eclipse without eye protection.

LUNAR ECLIPSE
During a lunar eclipse the Moon passes through Earth's shadow but it doesn't disappear from view. Instead it turns a dark reddish colour because red light can bend through Earth's atmosphere and reach it.

▶ **HIGH AND LOW TIDE**

The sea level at any coastline changes daily. On the Island of Mont St Michel off the coast of France, the difference between high and low tide is about 10 m (33 ft) but can reach as much as 16 m (52 ft) during a spring tide. At low tide visitors can walk to the island, but at high tide it is cut off by the sea, which made it a natural fortress in the past.

High tide

HOW TIDES WORK

The Sun and Moon both pull on Earth through their gravity, playing a constant game of tug-of-war with our planet. Although the Moon is much smaller than the Sun, its gravitational pull on Earth is twice as strong because it is closer. The combined forces drag the ocean around, creating the tides.

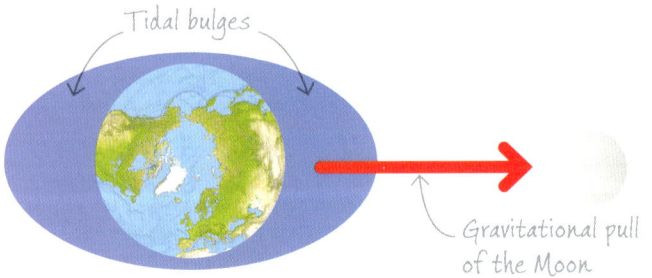

Tidal bulges

Gravitational pull of the Moon

DOUBLE BULGE

There are two high and two low tides every day. This is because the Moon creates two bulges in the ocean. The tidal bulge nearest the moon is caused by its gravitational pull. The opposite bulge is caused by inertia (the resistance to motion). Earth and the oceans are rotating, but inertia makes the water try to move in a straight line. As a result, it bulges outwards where the Moon's gravity is weakest.

SPRING TIDE

At full moon and new moon – when the Sun, Moon, and Earth line up – the Sun's and Moon's gravity combine to create the larger ocean bulges. This causes especially high and low tides, called spring tides (solar tide on the Sun's side and lunar tide on the Moon's side).

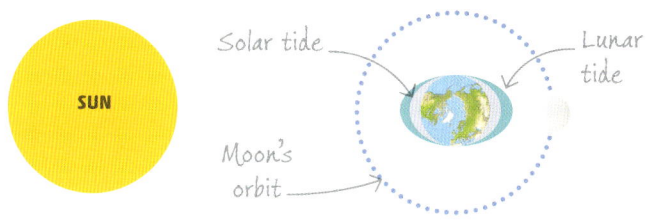

NEAP TIDE

When the Sun, Earth, and the Moon are at right angles to each other, the gravity of the Sun cancels out the gravity of the Moon, making the bulges smaller. This creates neap tides, when the difference between high tide and low tide is at its smallest.

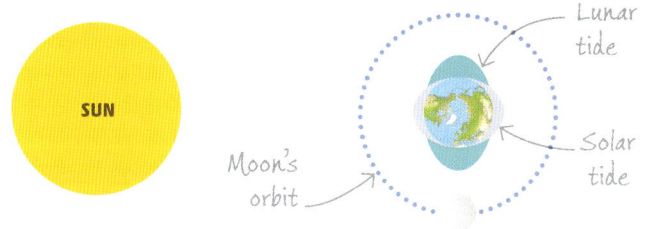

TIDAL BORE

In some parts of the world, the incoming tide is funnelled by a wide bay towards a river mouth, temporarily reversing the flow of the river and causing a powerful wave called a tidal bore. These rare waves create a powerful roar as the opposing currents crash and churn together. They attract surfers and sightseers but are occasionally deadly.

If you could cut Earth in half like an onion, you'd see a series of inner layers that get hotter and hotter with depth. The surface layer is a skin of cold, brittle rock: the **crust**. Under this is a **mantle** of red-hot but solid rock and a partly molten **core** made of metal. Heat from the interior makes Earth's outer layers move, slowly but continually changing the shapes of **continents** and oceans.

INSIDE THE PLANET

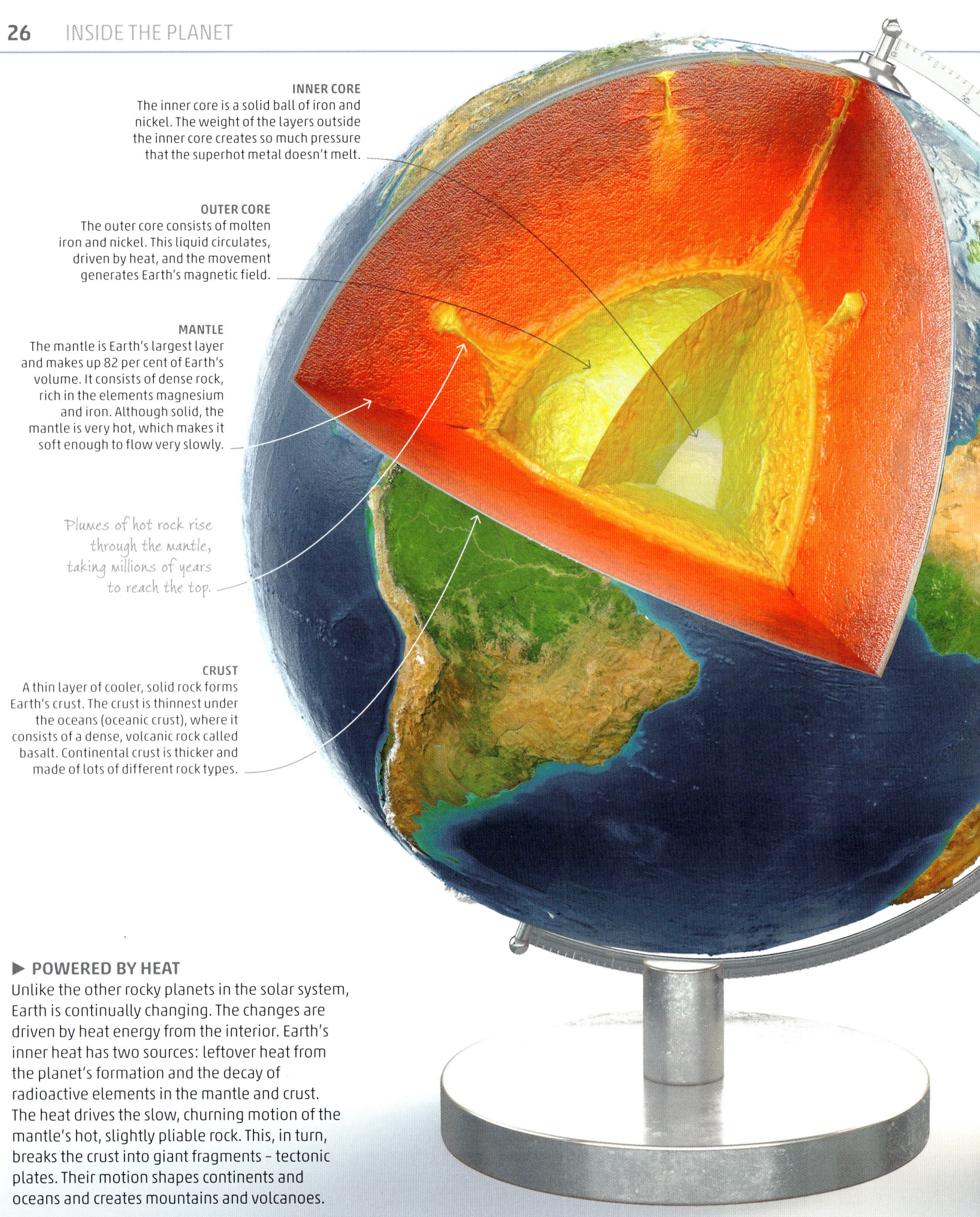

INNER CORE
The inner core is a solid ball of iron and nickel. The weight of the layers outside the inner core creates so much pressure that the superhot metal doesn't melt.

OUTER CORE
The outer core consists of molten iron and nickel. This liquid circulates, driven by heat, and the movement generates Earth's magnetic field.

MANTLE
The mantle is Earth's largest layer and makes up 82 per cent of Earth's volume. It consists of dense rock, rich in the elements magnesium and iron. Although solid, the mantle is very hot, which makes it soft enough to flow very slowly.

Plumes of hot rock rise through the mantle, taking millions of years to reach the top.

CRUST
A thin layer of cooler, solid rock forms Earth's crust. The crust is thinnest under the oceans (oceanic crust), where it consists of a dense, volcanic rock called basalt. Continental crust is thicker and made of lots of different rock types.

▶ POWERED BY HEAT
Unlike the other rocky planets in the solar system, Earth is continually changing. The changes are driven by heat energy from the interior. Earth's inner heat has two sources: leftover heat from the planet's formation and the decay of radioactive elements in the mantle and crust. The heat drives the slow, churning motion of the mantle's hot, slightly pliable rock. This, in turn, breaks the crust into giant fragments – tectonic plates. Their motion shapes continents and oceans and creates mountains and volcanoes.

WHAT'S INSIDE EARTH?

Early in Earth's history, the planet was so hot that it was almost completely molten. Heavy, dense elements such as iron and nickel sank into the centre to form a core. Lighter, molten rock, rich in the elements oxygen, silicon, and aluminium, rose to form a mantle and crust. And so the young planet separated into layers that still exist today.

Continental crust didn't start forming until Earth was over half a billion years old. The first crust was the solidified surface of the mantle and no longer exists. The crust we have today formed later, when magma rose beneath volcanoes and solidified to form rocks less dense than the mantle.

RISING TEMPERATURE

The deeper you go into Earth, the hotter it gets. The temperature rises about 25°C per 1 km (1°F per 70 ft) until you reach the inner core, where it's about as hot as the surface of the Sun. Throughout the core and lower mantle, rock and metal are white-hot and emit dazzling light.

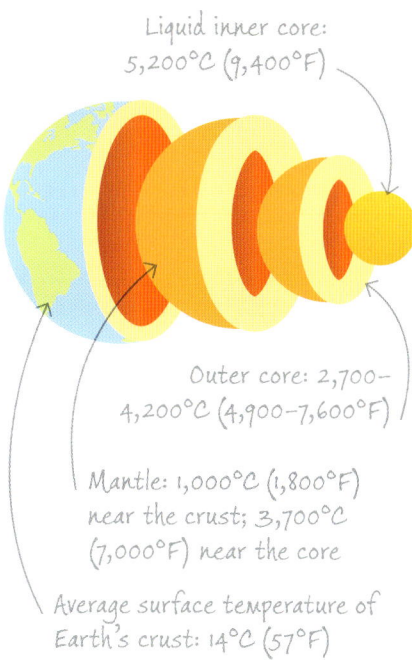

Liquid inner core: 5,200°C (9,400°F)

Outer core: 2,700–4,200°C (4,900–7,600°F)

Mantle: 1,000°C (1,800°F) near the crust; 3,700°C (7,000°F) near the core

Average surface temperature of Earth's crust: 14°C (57°F)

WATER WORLD

Earth is the only planet in the solar system with surface water in all three states: solid (ice), liquid, and gas (water vapour). There is also water deep inside the planet, where it exists as ions (charged particles) attached to minerals.

All Earth's surface water would make a sphere one ninth as wide as Earth.

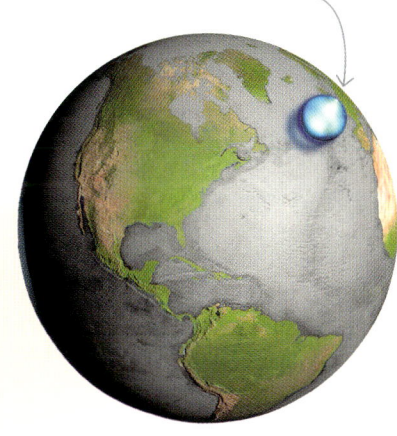

ROCKS FROM THE DEEP
Although we can't see the mantle directly, fragments of it are sometimes brought to the surface by volcanic eruptions. Mantle xenoliths are rocks from the upper mantle. They reveal that most of the upper mantle is a dense, grainy rock made of two minerals: olivine, which forms green crystals, and pyroxene, which is black.

HOW EARTH'S MAGNETISM WORKS

▼ **MAGNETIC FIELD**
The magnetic field around a magnet is the area where magnetic materials, such as iron, respond to the pull of a magnetic force. A magnetic field is normally invisible, but we can see the field around a magnet by sprinkling iron filings onto it. These tiny flecks of iron arrange themselves along lines of force, showing the direction in which the field pulls them.

The swirling movement of molten metal in Earth's outer core creates a magnetic field around the planet through a process called the dynamo effect. This field is shaped like that of a bar magnet, but on a much larger scale. Earth's magnetic field shields us from dangerous cosmic radiation, and it makes compass needles point north. The field's direction is recorded in certain rocks as they form. Scientists can read this record to understand how the world has changed over time.

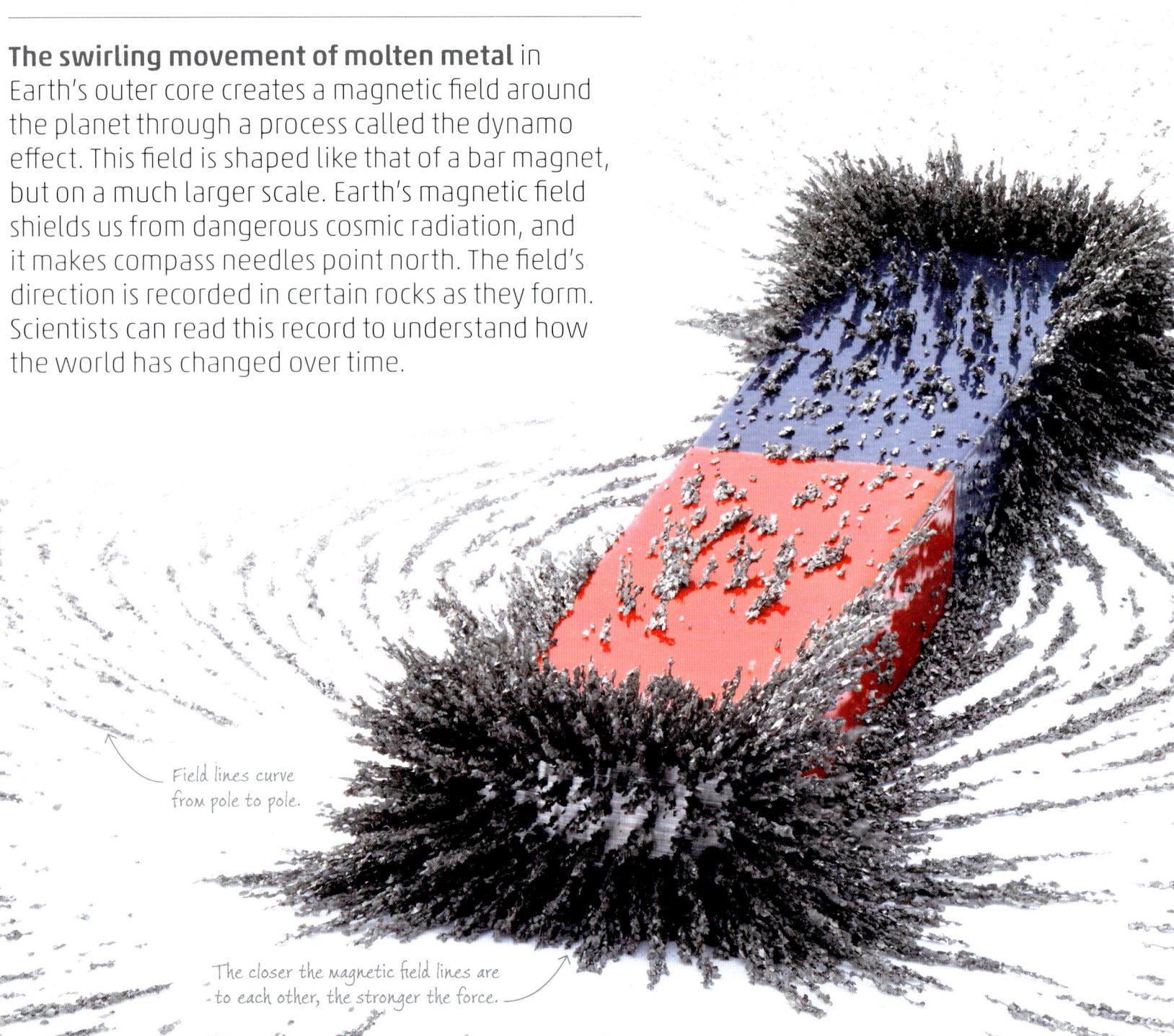

Field lines curve from pole to pole.

The closer the magnetic field lines are to each other, the stronger the force.

EARTH'S FIELD

Earth's magnetic field is shaped as though the planet contains a gigantic bar magnet. However, Earth's field is more complex. It isn't perfectly symmetrical and it's currently tilted about 11° from the axis of rotation, which means the geographic poles are not in the same place as the magnetic poles.

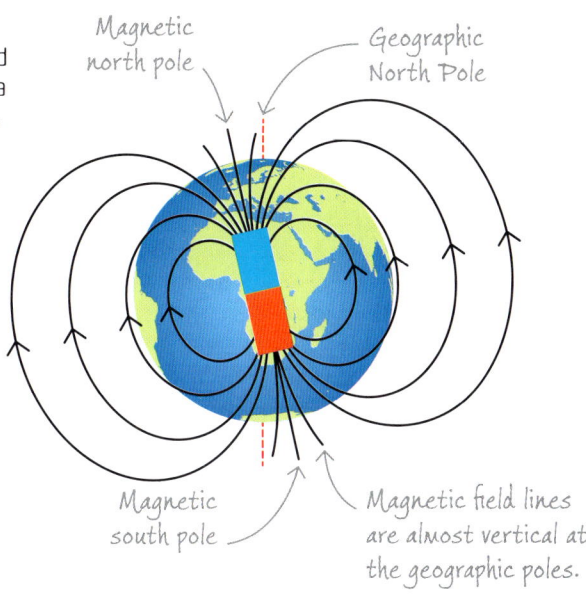

Magnetic north pole

Geographic North Pole

Magnetic south pole

Magnetic field lines are almost vertical at the geographic poles.

Iron filings

WANDERING POLES

The movement of Earth's molten core causes the magnetic poles to wander about randomly over time. The north and south magnetic poles even swap position every few hundred thousand years. Earth's magnetic north pole currently works like the south pole of a magnet, since it attracts the north pole of compasses (opposite poles attract).

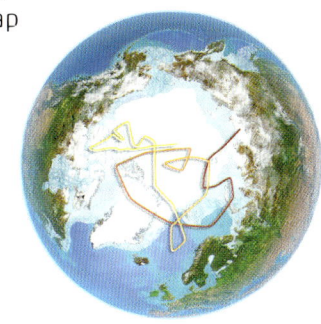

MAGNETIC NORTH POLE OVER THE PAST 2000 YEARS

PALEOMAGNETISM

When molten rock cools and hardens, minerals such as magnetite align with Earth's magnetic field. These patterns stay preserved in the rock crystals, and can tell scientists where on Earth the rock was formed. This branch of science is called paleomagnetism.

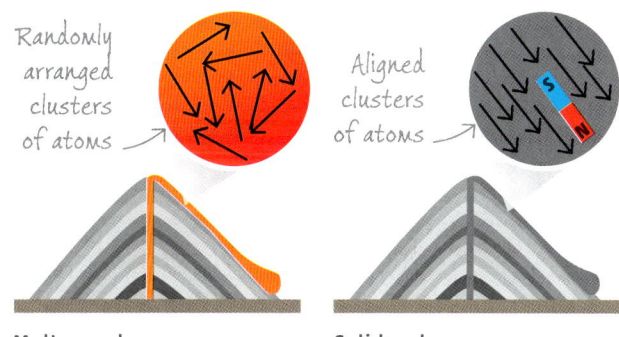

Randomly arranged clusters of atoms

Aligned clusters of atoms

Molten rock
When rock is molten, the tiny magnetic fields around atoms are arranged randomly.

Solid rock
As rock solidifies and crystals from, clusters of atoms become aligned with Earth's magnetic field.

PLATE TECTONICS

Paleomagnetism helped scientists to confirm the theory of plate tectonics. They discovered that rocks around a diverging plate boundary had a symmetrical pattern, with bands of rock displaying alternating magnetic polarity. These formed over a long period of time, as the plates moved apart and Earth's magnetic field flipped repeatedly.

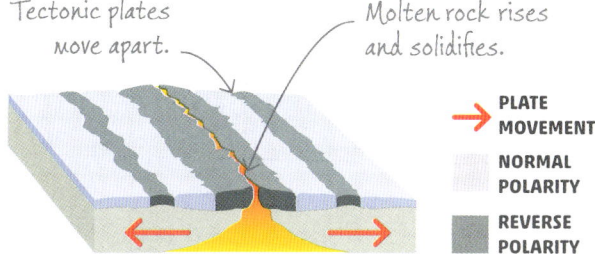

Tectonic plates move apart.

Molten rock rises and solidifies.

PLATE MOVEMENT

NORMAL POLARITY

REVERSE POLARITY

AURORA COLOURS

The colours in the aurora come from different elements in Earth's atmosphere. The most common colour is green, which comes from oxygen atoms 100–300 km (60–190 miles) high. Oxygen higher than this emits red light, and nitrogen atoms give off blue and purple. Mixtures of colours occasionally produce other hues, including yellow and pink.

STORMS ON THE SUN

The strongest and most colouful auroras are seen after coronal mass ejections – occasional eruptions of matter from the Sun. These solar storms hurl vast quantities of energized particles towards Earth. As well as producing brilliant auroras, they sometimes damage satellites and GPS systems.

AURORAL ZONES

Auroras are seen most often in two rings about 5,000 km (3,000 miles) wide around the poles. The northern aurora is called the aurora borealis (northern lights). The southern aurora is called the aurora australis (southern lights). After a large solar storm, auroras can occasionally be seen further away too.

AURORA AUSTRALIS

HOW THE
AURORA
WORKS

Visit Earth's polar regions in winter

and you might be lucky enough to see the world's greatest natural light show. The aurora lights up the night sky with shimmering veils of colour that pulse and wave from minute to minute. It is caused by the solar wind – a stream of charged particles from the Sun – colliding with gas atoms high in Earth's atmosphere.

This charged metal ball represents the Sun.

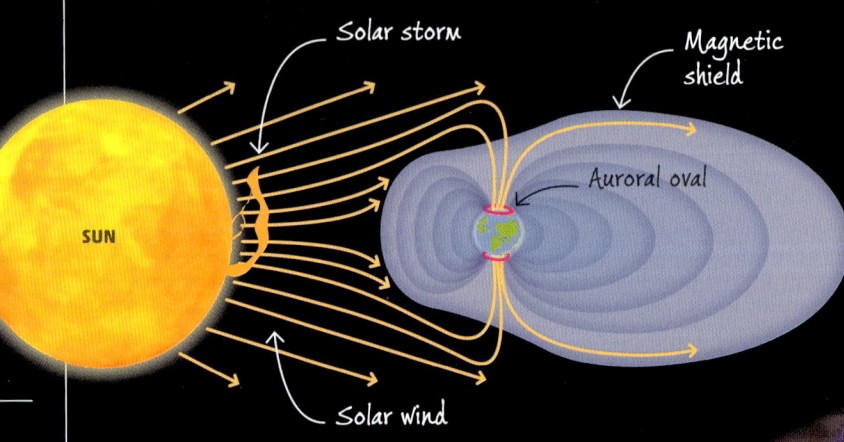

Solar storm

Magnetic shield

Auroral oval

SUN

Solar wind

THE MAGNETOSPHERE

Earth's magnetic field acts as a kind of shield, protecting the planet's surface from the solar wind. However, some of the solar particles get through. They are funnelled by the magnetic field towards the North and South Poles, where they collide with gas atoms in the atmosphere and create the auroras.

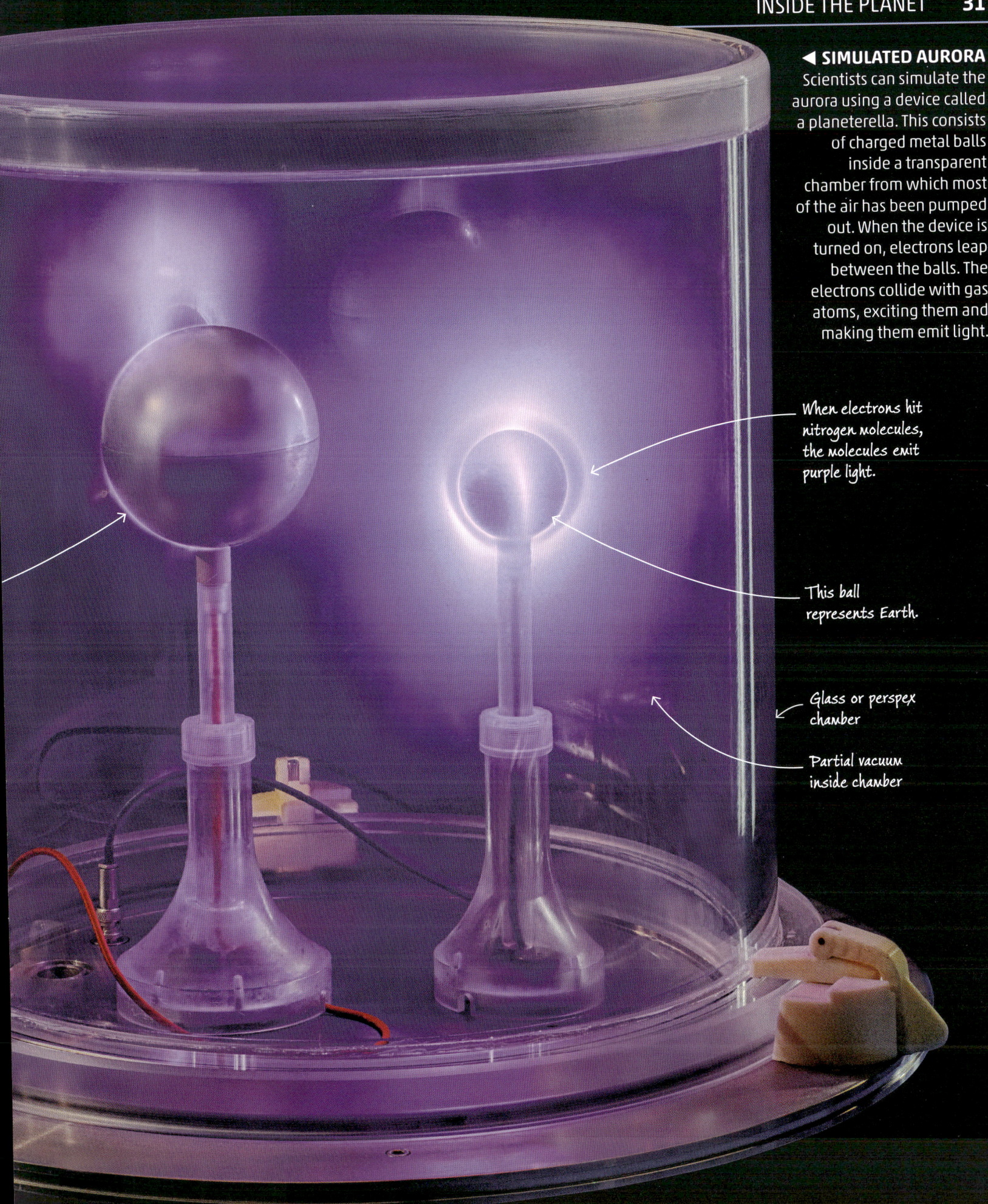

◀ **SIMULATED AURORA**
Scientists can simulate the aurora using a device called a planeterella. This consists of charged metal balls inside a transparent chamber from which most of the air has been pumped out. When the device is turned on, electrons leap between the balls. The electrons collide with gas atoms, exciting them and making them emit light.

When electrons hit nitrogen molecules, the molecules emit purple light.

This ball represents Earth.

Glass or perspex chamber

Partial vacuum inside chamber

HOW TECTONIC PLATES WORK

Earth's surface is broken up into 15 gigantic jigsaw pieces called tectonic plates. These move very slowly – just a few centimetres a year, which is about as fast your toenails grow. In some places they collide head-on, creating mountains and causing earthquakes and volcanic eruptions. In other places they move apart, with new crust forming between them.

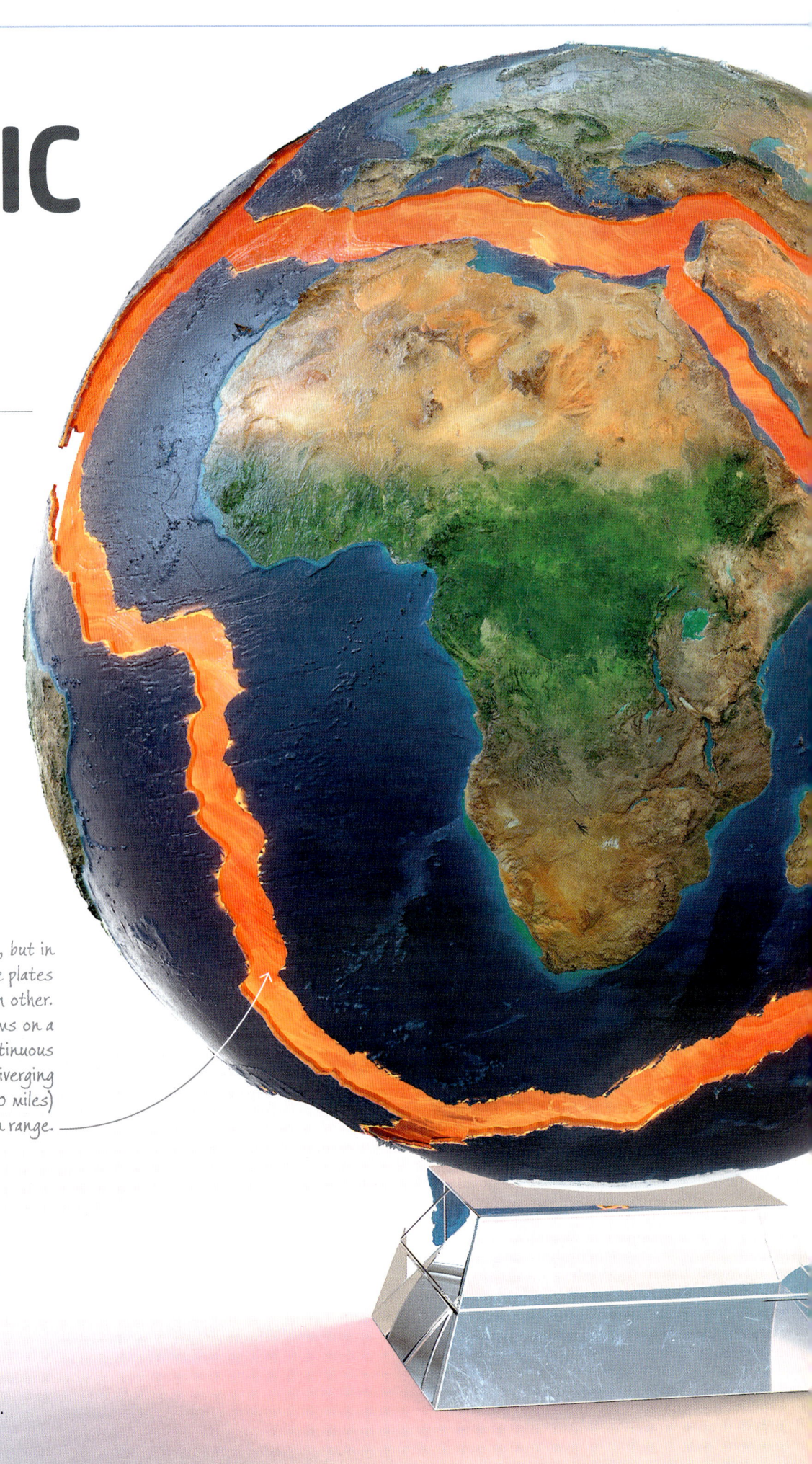

The plates are separated in this model, but in reality they meet at boundaries where plates either converge, diverge, or slide past each other. Stretching around the planet like seams on a baseball is the mid-ocean ridge – a continuous range of underwater volcanoes along diverging plate boundaries. At 65,000 km (40,400 miles) long, it is the world's longest mountain range.

▶ **AFRICAN PLATE**

The African plate is one of the largest tectonic plates and includes part of the Atlantic seafloor as well as the continent of Africa. This plate is beginning to split into separate plates due to a rift running down the mountains of eastern Africa. Millions of years from now, a new ocean will form along this rift, dividing the continent in two.

Arabian plate

The Pacific and North American plates are grinding past each other as they slide in opposite directions.

At the Californian coast the tectonic plate is moving about 5 cm (2 in) a year, which is much faster than most plates.

Nearly all of the Pacific Ocean is part of the Pacific plate, the largest of all the plates.

The Pacific and Nazca plates are moving away from each other.

The Nazca plate is moving towards and under the South American plate.

Indian plate

Antarctic plate

NORTH AMERICAN PLATE

JUAN DE FUCA PLATE

EURASIAN PLATE

ARABIAN PLATE

INDIAN PLATE

CARIBBEAN PLATE

PACIFIC PLATE

COCOS PLATE

PHILIPPINE PLATE

NAZCA PLATE

SOUTH AMERICAN PLATE

AFRICAN PLATE

AUSTRALIAN PLATE

ANTARCTIC PLATE

SCOTIA PLATE

EARTH'S TECTONIC PLATES
There are seven major tectonic plates: the North American, South American, Pacific, African, Eurasian, Australian, and Antarctic plates. There are at least eight other minor plates. They fit together snugly at plate boundaries.

THE LITHOSPHERE
Tectonic plates are made of more than just Earth's crust. They also include the upper part of the mantle. Together the crust and upper mantle form a cool and very rigid layer called the lithosphere. Under the lithosphere, the mantle rock is hotter and almost at melting temperature. This rock is softer and flows very slowly, carrying tectonic plates along with it.

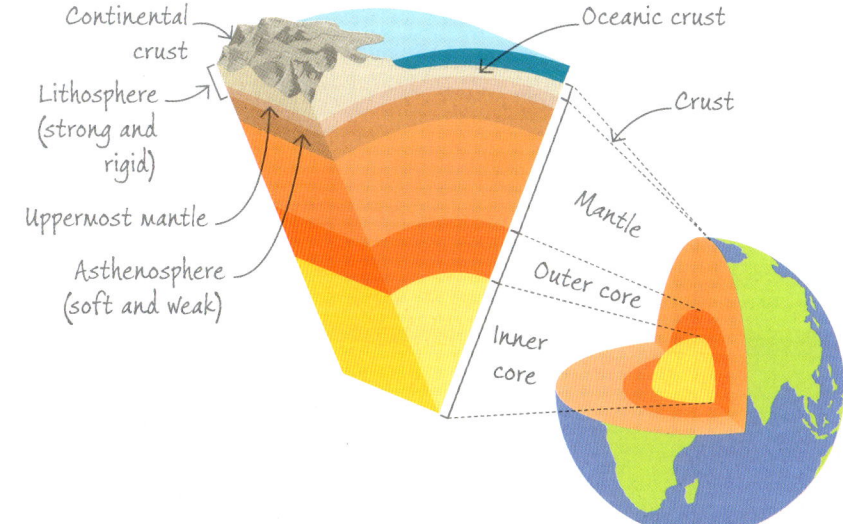

Continental crust

Oceanic crust

Lithosphere (strong and rigid)

Crust

Uppermost mantle

Mantle

Asthenosphere (soft and weak)

Outer core

Inner core

A TECTONIC DISCOVERY
US geologist Marie Tharp (1920–2006) was one of the scientists who discovered the mid-ocean ridge system, laying the foundations for the revolutionary theory of plate tectonics. Tharp and her colleague Bruce Heezen used depth measurements from ships to map the Atlantic seafloor. Their map revealed a huge mountain chain and rift valley running down the middle of the ocean.

TORN APART

In some parts of the world, tectonic plates are moving apart from each other, tearing open giant cracks and canyons in Earth's crust. The country of Iceland sits directly on the boundary between the North American and Eurasian tectonic plates, which are pulling apart. Such rifts are usually hidden below the sea, but in Iceland the tectonic scars are visible at the surface.

▶ CONVECTION IN A LAVA LAMP

Turn on a lava lamp to see convection in action. A bulb in the base heats up paraffin wax surrounded by liquid. The hot wax expands and so becomes less dense. This makes it buoyant, so it rises, expanding more as it goes. As it gets further away from the heat, it cools and becomes denser than the surrounding liquid, which makes it sink due to the pull of gravity. A similar process occurs in Earth's mantle.

Paraffin wax expands as it heats.

The wax rises as it expands.

The wax will start to cool as it moves away from the heat source.

A bulb in the base releases heat.

HOW TECTONIC PLATES MOVE

The tectonic plates that make up Earth's outermost layer are continually on the move, inching along at about the same speed that human toenails grow. This motion is driven by processes deep inside the planet. The details are not yet fully understood by science, but the most important drivers of tectonic plates are convection and the rise of hot, soft rock in Earth's mantle.

MANTLE CONVECTION

Although Earth's mantle is mostly solid rock, it is softened by heat from the core. Over millions of years, softened rock rises by convection like paraffin wax in a lava lamp. This transfer of heat energy from the core to the crust drives the motion of tectonic plates. However, convection inside Earth is more complex than in a lava lamp: the mantle might be divided into multiple layers, and the brittle crust does not flow like a liquid.

In some parts of the mantle, hot rock rises towards the surface in plumes like the hot wax in a lava lamp.

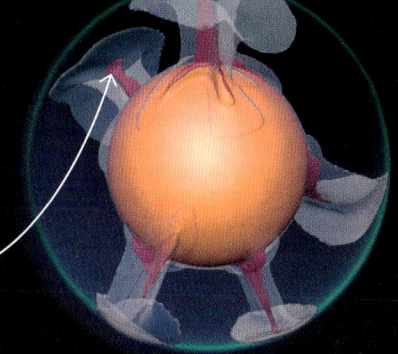

COMPUTER MODEL OF MANTLE CONVECTION

RIVAL THEORIES

Scientists have proposed two different models for mantle convection, but they aren't sure which is right.

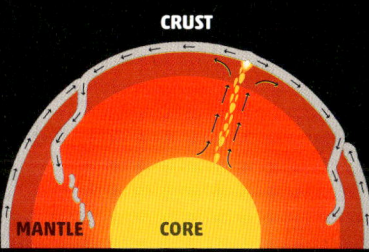

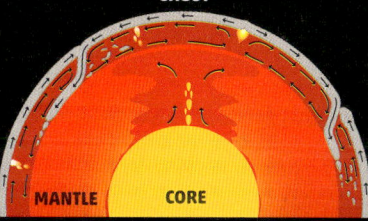

WHOLE MANTLE CONVECTION

According to this model, the whole mantle is stirred by convection. Plumes of hot rock rise from the core to the crust, where they cool, create new crust, and push tectonic plates apart. In other places, plates collide and can sink to the bottom of the mantle.

LAYER CAKE MODEL

In the layer cake model, the mantle is divided into different layers, each with its own convection cycle. The sinking tectonic plates stop at the boundary between the upper and lower mantle, and the plumes that push plates apart form at shallow depths.

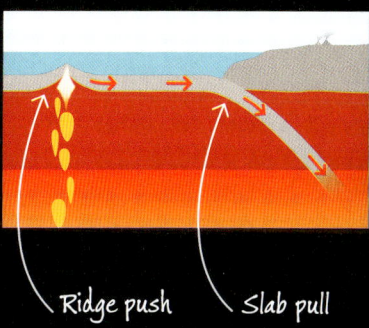

Ridge push   Slab pull

THE POWER OF GRAVITY

Gravity plays an important role in moving tectonic plates. At mid-ocean ridges, hot parts of the mantle rise and lift the ocean floor. As new crust forms at the ridge, it pushes the crust sideways and gravity pulls it down (ridge push). Gravity also pulls down the subducting (sinking) edges of tectonic plates, as the rock here is more dense and heavy than the surrounding mantle (slab pull).

HOW PLATES COLLIDE

The tectonic plates that make up Earth's crust meet at tectonic boundaries. At some boundaries, one plate slides under the other. At others, the plates move apart or they grind past each other horizontally. These three types of plate boundary are called convergent, divergent, and transform.

▼ COLLISION ZONES
Tectonic boundaries can occur on land, under the sea, and where land meets sea. These meeting points are the most dynamic places on the planet. Here, mountain ranges rise as the crust buckles and folds. In the depths of the sea, giant trenches form where weaker oceanic crust is forced down (subducted) into the planet's interior.

CONVERGENT BOUNDARIES
Convergent boundaries are where two plates collide head-on. In a collision between continental and oceanic crust, the denser oceanic plate sinks below its lighter neighbour. This is called subduction. The subducting plate carries water into the mantle, which causes rock to melt and volcanoes to form. Meanwhile, the continental crust folds and buckles, creating mountains.

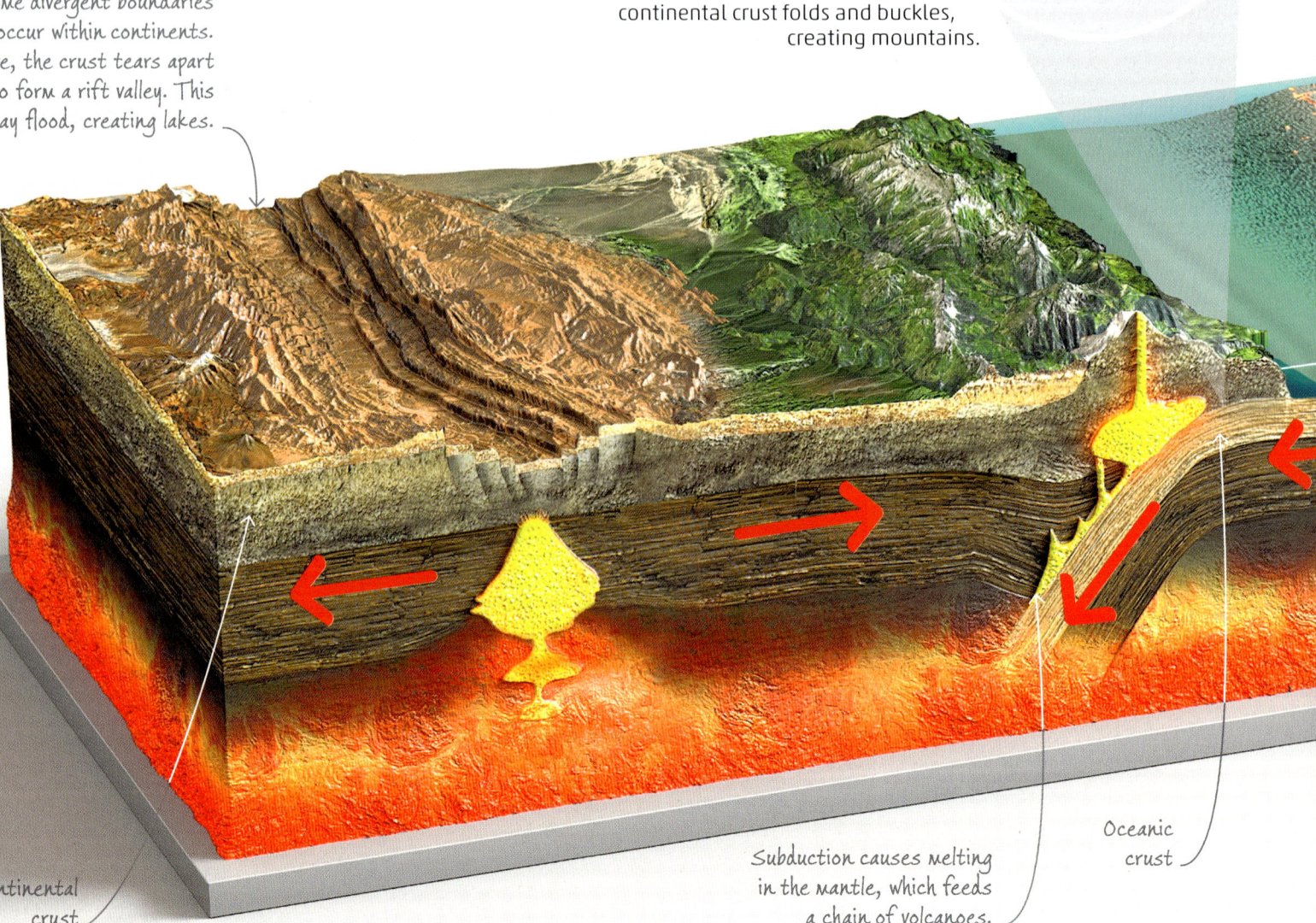

Some divergent boundaries occur within continents. Here, the crust tears apart to form a rift valley. This may flood, creating lakes.

Subduction causes melting in the mantle, which feeds a chain of volcanoes.

Oceanic crust

Continental crust

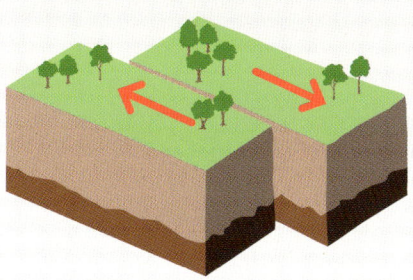

TRANSFORM BOUNDARIES
Transform plate boundaries are where neighbouring plates move parallel, grinding past each other. Most transform faults are on the sea floor, but some are on land.

LAND DIVIDED
The San Andreas fault is a transform boundary that cuts across California. On average, the two plates move past each other by only 2.5 cm (1 in) a year. However, the movement occurs in fits and starts, generating powerful earthquakes.

HOT SPRINGS
Geothermal features such as hot springs, geysers, and boiling mud pots are common at tectonic boundaries. These all form when groundwater is heated by magma that has risen into the crust.

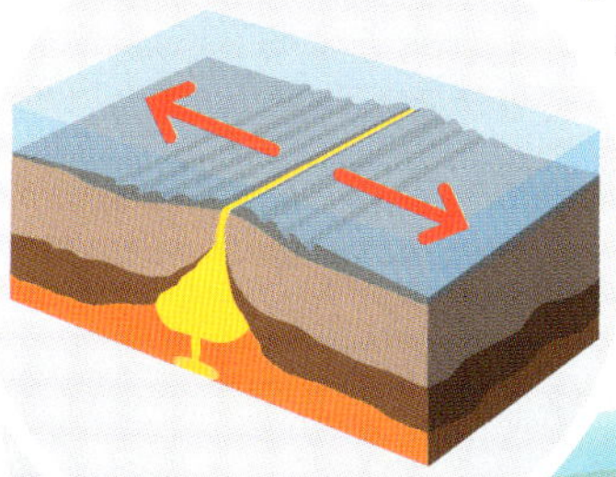

DIVERGENT BOUNDARIES
Divergent plate boundaries are where neighbouring plates move apart (diverge). As the plates shift, long cracks develop. Hot rock from the mantle rises, partially melts, and fills the cracks. This very slowly creates new crust.

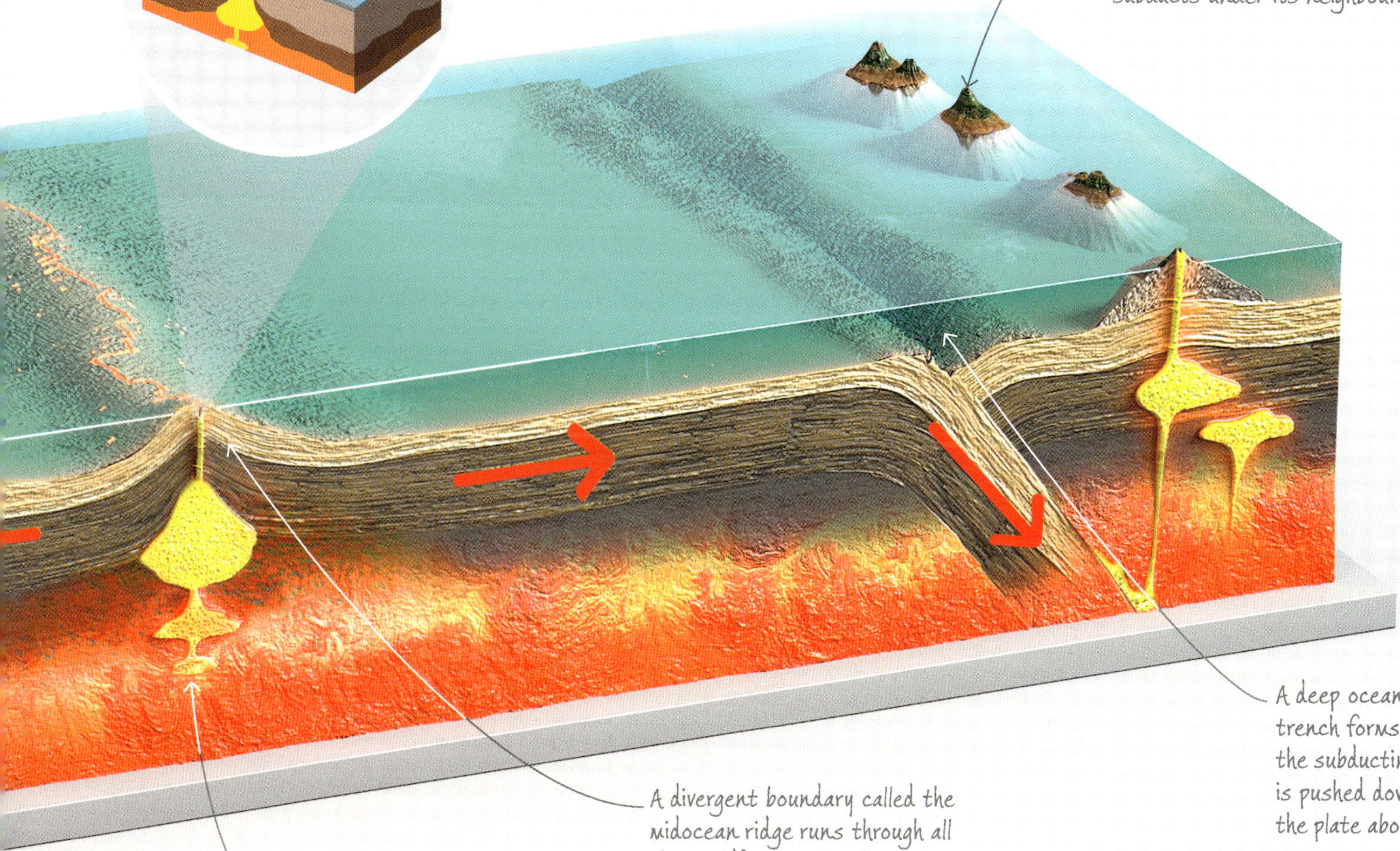

An arc of volcanic islands forms near an ocean–ocean convergent boundary, where one oceanic plate subducts under its neighbour.

A deep oceanic trench forms where the subducting plate is pushed down by the plate above.

A divergent boundary called the midocean ridge runs through all the world's oceans. The sea floor is higher here, forming the world's longest mountain chain.

The upper mantle partially melts under divergent boundaries.

360 MILLION YEARS AGO

Laurentia and Baltica combined to form the supercontinent of Laurasia. This was separated from Gondwana by sea, but the two supercontinents were heading towards each other on a collision course.

LAURASIA

GONDWANA

420 MILLION YEARS AGO

The continents of Baltica (including parts of modern Europe) and Laurentia (North America and Greenland) moved together. The ocean between them slowly shrank until it was gone.

LAURENTIA **BALTICA**

500 MILLION YEARS AGO

Half a billion years ago, Earth's northern hemisphere was mostly ocean. A supercontinent called Gondwana, along with several smaller continents, lay in the southern hemisphere.

GONDWANA

PANTHALASSA OCEAN

PANGAEA

MOUNTAINS

300 MILLION YEARS AGO

Laurasia and Gondwana collided to form a single, vast supercontinent: Pangaea. The collision created a mountain range across its centre, stretching from present-day Mexico to Poland. A huge ocean, Panthalassa, surrounded Pangaea.

NORTH AMERICA

ATLANTIC

AFRICA

180 MILLION YEARS AGO

A rift tore Pangea into two tectonic plates. The rift valley flooded, forming the beginnings of the North Atlantic Ocean between what are now North America and Africa.

NORTH AMERICA

NORTH ATLANTIC

AFRICA

SOUTH AMERICA

INDIA

ANTARCTICA

120 MILLION YEARS AGO

More rifting caused the South Atlantic Ocean to form, opening like a zip to separate South America from Africa. India and Antarctica broke away from Africa and drifted slowly away.

CONTINENTAL DRIFT

The theory that continents move was put forward by German scientist Alfred Wegener in 1912. He noticed that South America and Africa fit like jigsaw pieces and suggested they were once joined. He found matching rocks and fossils on each continent, but he couldn't explain what had pushed them apart, and his idea was ridiculed. It wasn't until after he died attempting to cross Greenland's ice sheet that this theory was finally accepted.

HOW CONTINENTS CHANGE

Moving at about the speed that toenails grow, Earth's continents have shifted over time, carried by the moving tectonic plates that make up the planet's crust. Continents have merged into supercontinents and broken apart. Oceans have opened and closed. The evidence for these incredible changes comes from many sources, including fossils, sea floor surveys, and magnetic patterns in rocks.

THE WALLACE LINE

The Wallace Line is an imaginary line separating Asia from Australia and New Guinea. It marks the boundary between the ancient continents that formed when the Pangaea supercontinent broke up. After this separation, animals on either side of the line evolved in different ways. Marsupials such as kangaroos evolved to the east of the line, while mammals without pouches evolved to the west.

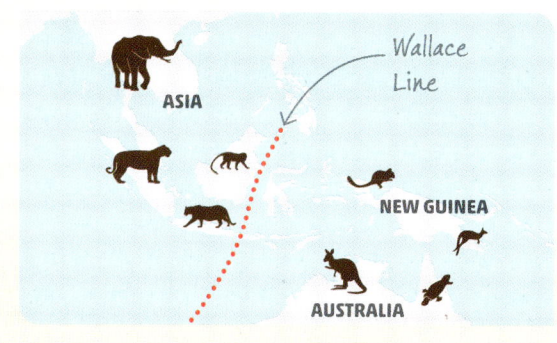

40 MILLION YEARS AGO
The Atlantic Ocean widened, pushing North America and Eurasia further apart, and the continents began to take on their modern shapes. Africa drifted north, closing the Tethys Ocean, until it crashed into Eurasia. The collision created the Alps mountains.

TODAY
Earth now has seven named continents rather than a single supercontinent. India is currently crashing into Eurasia, pushing up the Himalayas. The Pacific Ocean, although the world's biggest ocean, is shrinking.

The ground below your feet may feel rock solid, but the jigsaw of **tectonic plates** that make up Earth's **crust** is continually moving and cracking. Sudden jolts can shake the ground with great violence, causing **earthquakes**. Magma – liquid rock – oozes up through the crust from molten parts of the **mantle**, occasionally bursting out at the surface in **volcanic eruptions**.

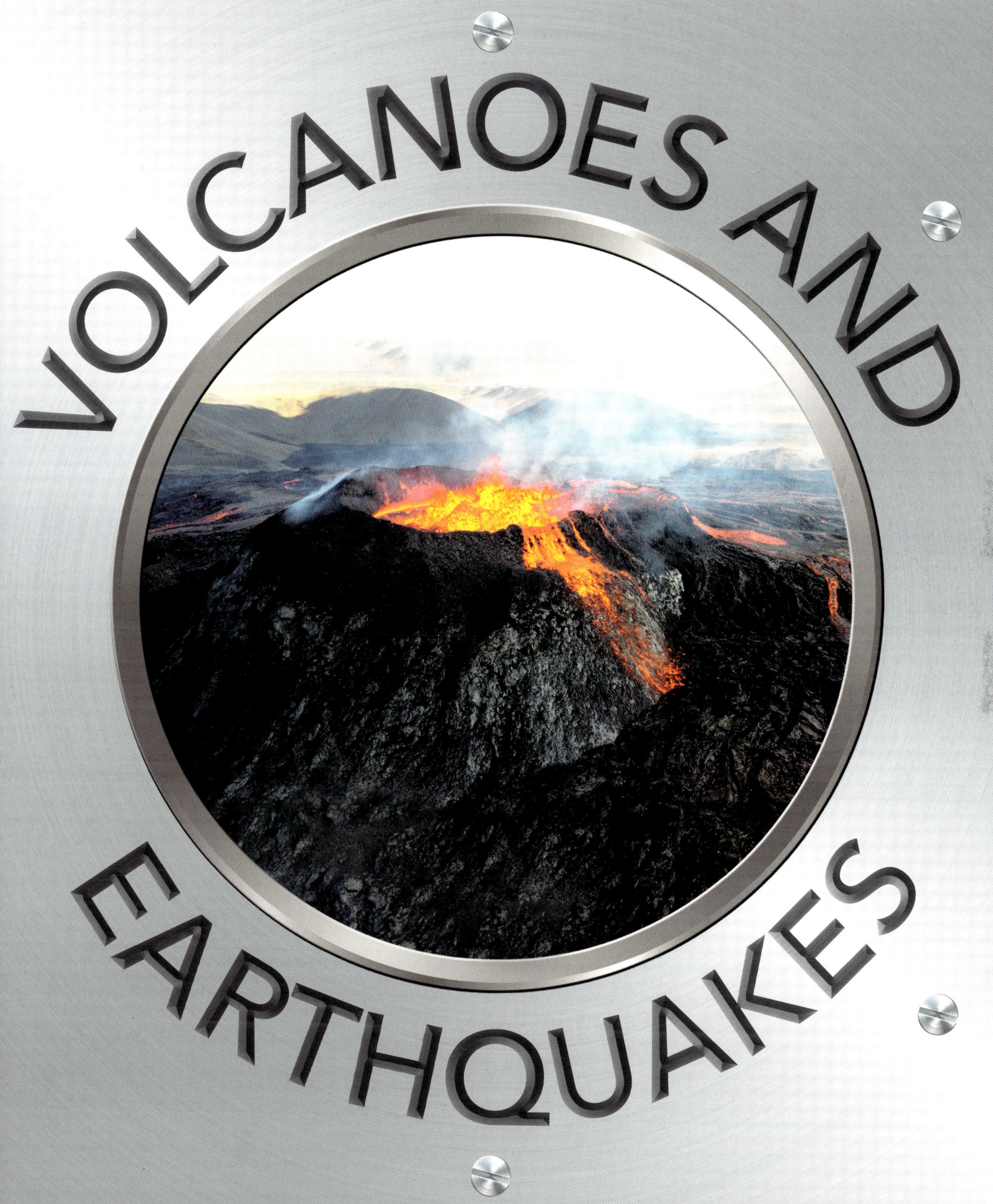

VOLCANOES AND EARTHQUAKES

VOLCANIC WORLD

Volcanoes and earthquakes are powerful reminders of the tectonic activity that takes place beneath our feet. We know where these deadly hazards are most likely to occur, but predicting when disaster will strike is very difficult.

The Aleutian Trench was formed by the Pacific plate sinking beneath the North American plate.

The Great Rift Valley is slowly splitting Africa apart.

EURASIAN PLATE

ARABIAN PLATE

AFRICAN PLATE

INDIAN PLATE

PHILIPPINE PLATE

INDO-AUSTRALIAN PLATE

ANTARCTIC PLATE

MOUNT VESUVIUS
In 79 CE Mount Vesuvius erupted in Italy, burying the city of Pompeii under ash and rock. Most of the people died from the extreme heat of the scalding ash.

TŌHOKU TSUNAMI
In 2011, an earthquake in the Pacific Ocean triggered a tsunami that slammed into northeastern Japan, destroying thousands of homes and causing a nuclear disaster at the Fukushima power plant.

▼ DYNAMIC PLANET

Most volcanoes and earthquakes happen at the boundaries between tectonic plates. In these collision zones, magma forms where plates push into each other or tear apart, and the molten rock oozes into the crust to feed eruptions. Earthquakes happen when plates or parts of plates move past each other in sudden jolts, sending shockwaves through the ground.

● **Volcanoes active in the last 10,000 years**

● **Earthquakes above magnitude 6 in the last 100 years**

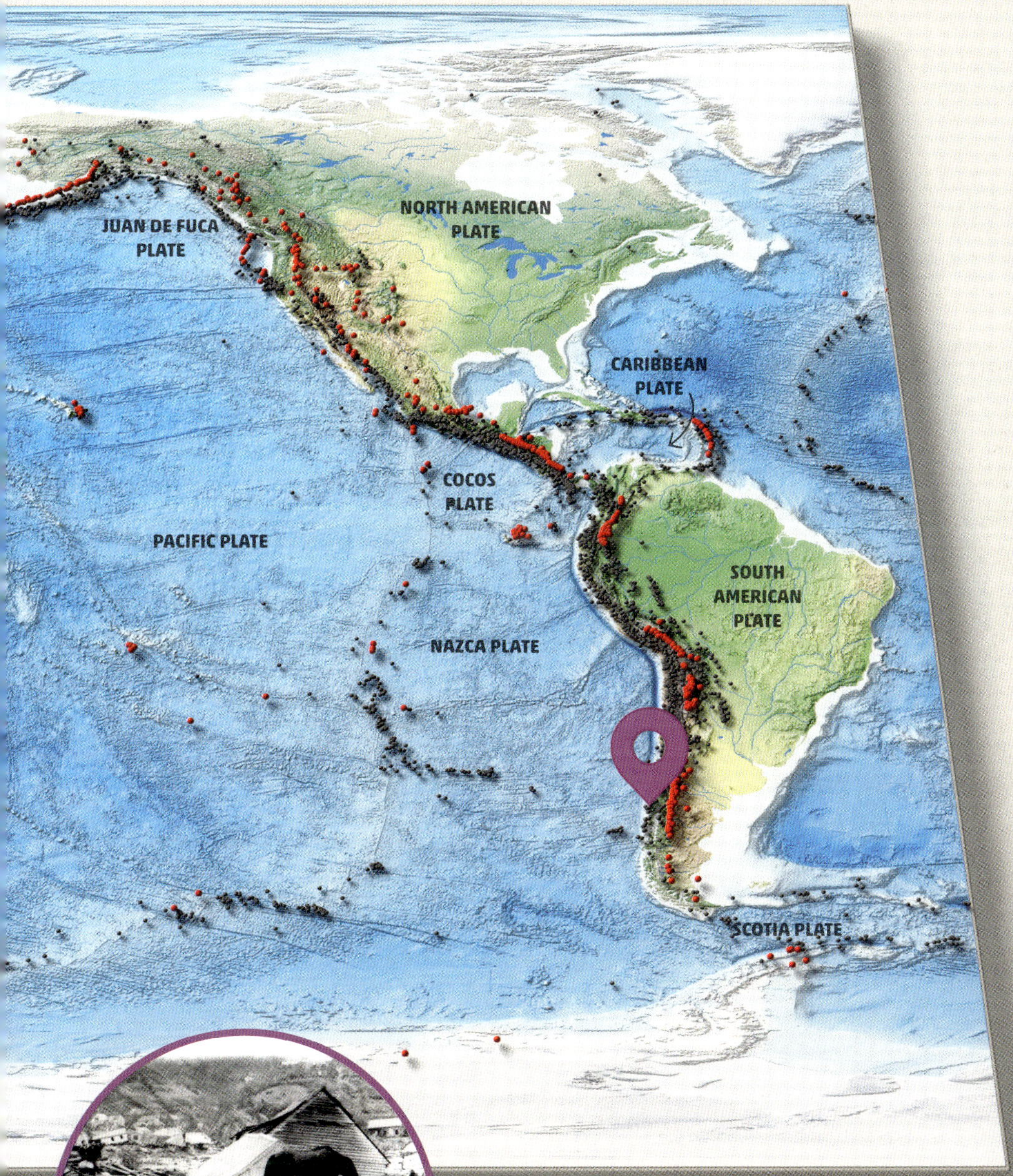

JUAN DE FUCA PLATE

NORTH AMERICAN PLATE

CARIBBEAN PLATE

COCOS PLATE

PACIFIC PLATE

NAZCA PLATE

SOUTH AMERICAN PLATE

SCOTIA PLATE

📍 VALDIVIA EARTHQUAKE
In 1960, the most powerful earthquake ever recorded struck near Valdivia, Chile, measuring a whopping 9.5 in magnitude. The quake triggered tsunamis, destroyed buildings, and left over 2 million people homeless.

RING OF FIRE

The Pacific Ring of Fire is home to three-quarters of Earth's volcanoes and is where 90 per cent of earthquakes happen. It is made up of several different plate boundaries and stretches from New Zealand to Russia and around the western coastlines of North and South America.

PACIFIC OCEAN

Ring of Fire →

ACTIVE VOLCANOES

Active volcanoes (volcanoes still connected to a magma chamber) occur in clusters. The USA has the most, with 165, whereas the continent of Australia has no volcanoes that have erupted in the last 1,000 years. About 500 volcanoes have erupted in the last century.

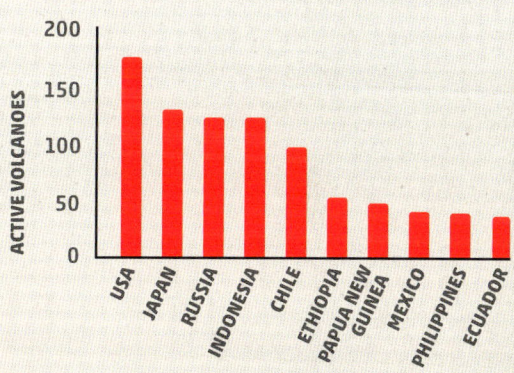

ACTIVE VOLCANOES

200 · 150 · 100 · 50 · 0

USA · JAPAN · RUSSIA · INDONESIA · CHILE · ETHIOPIA · PAPUA NEW GUINEA · MEXICO · PHILIPPINES · ECUADOR

THE 2-MILLION-YEAR ERUPTION

About 252 million years ago, 90 per cent of the world's species mysteriously died out. Some scientists think volcanoes were to blame: at around the same time, a massive volcano in Siberia produced floods of lava that kept flowing for 2 million years.

Layers of hardened lava and ash may build up to form a mountain.

Ash cloud

An extinct volcano has no magma supply and will never erupt again.

Magma is mostly molten but also contains solid rock crystals and bubbles of gas. Variations in the ratio of these affect whether a volcano erupts explosively or more calmly.

Shallow magma reservoir

Some magma pushes its way through the crust horizontally (forming sills), vertically (forming dykes), or at other angles.

Magma accumulates in chambers of various shapes and sizes.

Magma forms in the upper mantle and rises into the crust.

DEEP ROOTS
The roots of volcanoes can reach more than 100 km (62 miles) deep, extending all the way to the mantle. From there, a series of magma chambers and pathways bring molten rock to the surface in a journey that can take millennia. Inside magma chambers, rock crystals and gas bubbles form as the temperature and pressure vary. If the pressure gets too high, the magma is forced out in a volcanic eruption.

Some magma chambers contain a substance called crystal mush, which is not fully molten and is mostly made of solid crystals.

Melt pocket

HOW
VOLCANOES
WORK

When magma spills onto Earth's surface, we call it lava.

Volcanoes form when magma (molten rock) spills onto Earth's surface. Magma forms miles undergound in Earth's upper mantle. The mantle is normally solid, but small patches of melting happen at plate boundaries or other hot areas. The red-hot liquid then rises through the crust, seeping through cracks, melting through rock, and pooling in chambers. It can spend thousands of years in a magma chamber before finally erupting from a volcano.

Most lava erupts through the main conduit (vent) of a volcano.

VOLCANO SCIENCE

Volcanologists (volcano scientists) forecast eruptions by looking for signs that magma is on the move. Sometimes the ground rises ominously, lifted by a bulging magma chamber. Earthquakes may be detected as magma bursts through cracks underground. Rising magma may also release telltale gases such as carbon dioxide and sulphur dioxide, which can be measured.

CHANGING MAGMA

As magma is held in a magma chamber, its composition slowly changes as one element after another forms crystals and then sinks out of the liquid rock.

1 COOLING
Rock crystals form as magma cools. They are denser than liquid rock, so they sink and settle at the bottom.

2 New crystals
Crystallization uses up certain elements. After they run out, new kinds of crystal form.

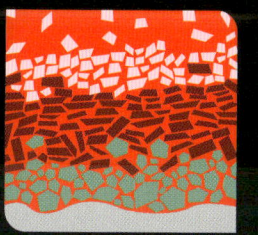

3 Magma changes
As the cycle continues, the chemical make-up of the magma changes. This affects how frequent and how dangerous any eruptions are.

▲ TUNGURAHUA

Tungurahua is one of Ecuador's most active volcanoes. Some of its eruptions produce rivers of lava, while others belch out ash clouds, rock fragments, and lava bombs. Layers of hardened lava and tephra (rock fragments) have piled up over thousands of years into a conical mountain, typical of this kind of volcano (a stratovolcano).

▶ **EXPLOSIVE ERUPTIONS**
On 18 May 1980, Mount St Helens in Washington State, USA, exploded in a Plinian eruption – the most explosive kind of eruption. An earthquake triggered a landslide that allowed pressurized magma and gas to explode northwards. The blast destroyed the summit and north face of the mountain, creating a crater 1.6 km (1 mile) wide and hurling 540 million tonnes of scalding ash into the sky.

1973

1982

HOW
VOLCANOES
ERUPT

Volcanic eruptions are among the most powerful natural events on Earth. They can blanket vast areas in ash, destroying towns and causing many deaths. But they also create new land and fertilize soil, which is good for farming. Every volcanic eruption is different. The type of eruption depends on the composition of the magma, how hot it is, how much gas it contains, and how sticky it is.

Crater

EFFUSIVE ERUPTIONS

Not all eruptions are explosive. When Russia's Tolbachik volcano burst into life in 2012, it produced rivers of runny lava that flowed for 20 km (12 miles). Eruptions of runny lava are called effusive eruptions and can continue for months. When the lava cools, it hardens to form a kind of rock called basalt.

Volcanic ash is a mixture of hot gases and tiny specks of rock and glass. The ash cloud can drift on the wind for miles before settling as a dusty layer on the ground.

PRESSURE RELEASE

Explosive eruptions are powered by bubbles. Gas bubbles form in magma as it sits in a magma chamber. In runny magma, bubbles rise to the surface and pop, but in sticky magma they build up. A sudden release of pressure makes the bubbles expand, turning magma to foam. The result is an explosion – a bit like shaking a fizzy drink and opening it.

ERUPTION STYLES

Volcanologists classify eruptions into several different types based on their size and how explosive they are.

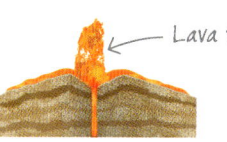

Lava fountain

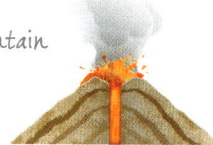

Hawaiian eruptions are the most effusive and are characterized by lava fountains and flows of runny, basaltic magma.

Strombolian eruptions are small explosions caused by gas bubbles in magma channels. They can throw lava bombs hundreds of metres high.

Pyroclastic flow

Vulcanian eruptions are short, explosive bursts that happen when sticky lava blocking a vent is suddenly blown out.

Pelean eruptions are powered by trapped gas and are explosive. They cause deadly avalanches of ash (pyroclastic flows).

Plinian eruptions are the biggest and deadliest. So much gas is trapped in the sticky magma that it turns to foam in an instant when pressure is released. The foam explodes outwards at terrific speed and immediately solidifies, breaking up into an ash cloud that can rise in a plume many miles high.

Ash plume

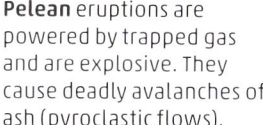

MAGMA AND WATER

When magma meets water, heat turns the liquid water into vapour, which can be explosive. In 1963, a massive eruption on the Atlantic sea floor created the island of Surtsey off the coast of Iceland. Such eruptions are now called Surtseyian after the island.

TYPES OF VOLCANO

Every volcano is different. Some volcanoes tower over the landscape as mountains, while others are just holes in the ground or hidden entirely underwater. The most active volcanoes spew out lava continually, but other volcanoes can stay dormant for centuries before exploding without warning. There are six main types of volcano: lava domes, fissure volcanoes, cinder cones, calderas, shield volcanoes, and stratovolcanoes.

LAVA DOME AT NOVARUPTA VOLCANO, ALASKA

LAVA DOMES
When viscous (sticky) lava erupts, it can't flow away. Instead it oozes out slowly and builds up into a steep-sided dome. Sometimes the inside of the dome stays molten and forces its way out as a spike. Or the dome may completely collapse, resulting in an avalanche of volcanic debris.

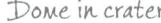
Dome in crater
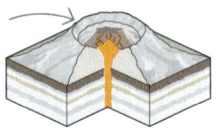

FISSURE ON KILAUEA VOLCANO, HAWAII

FISSURE VOLCANOES
A fissure volcano is a long crack in the ground through which runny lava erupts, often as a curtain. These eruptions feed some of the largest lava flows on Earth.

Curtain of lava
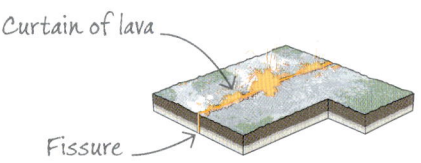
Fissure

CINDER CONES
These small, conical volcanoes are heaps of loose volcanic fragments that pile up when a fountain of lava erupts from a volcanic vent. They have a central crater and steep slopes that are easily eroded. They can grow on the side of stratovolcanoes and sometimes occur in clusters.

Crater

Steep cone made of loose debris

MEKE CRATER, TURKEY

CALDERAS

If a violent volcanic eruption empties part of a magma chamber just below the volcano, the ground may collapse to form a crater called a caldera. Calderas can be miles wide and often flood with water to form lakes.

Wide, water-filled crater

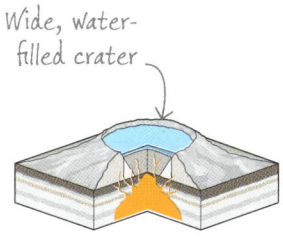

QUILOTOA LAKE , ECUADOR

SHIELD VOLCANOES

Shield volcanoes build up from runny lava that spreads out over a large area. They are very wide with gentle slopes but can grow enormous. They include the largest volcanoes on Earth.

Gentle slope

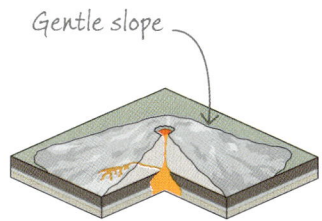

SUMMIT OF MAUNA LOA, HAWAII

STRATOVOLCANOES

These large volcanoes have a distinctive conical shape that makes them easy to recognize as volcanoes. They grow layer by layer from eruptions of thick, sticky lava that doesn't flow far. Between the lava flows are layers of ash and pumice from explosive eruptions.

Alternating layers of lava and ash

MOUNT FUJI, JAPAN

HOW **LAVA** FLOWS

Lava is molten rock that flows onto Earth's surface from a volcano. Not all lava is the red-hot liquid we are familiar with. Depending on its temperature and composition, lava can be as runny as syrup, thick and gloopy like porridge, or so stiff that it resembles a sliding heap of rubble. Even thick lava can flow a surprisingly long way, while its molten interior stays insulated by a solid outer crust.

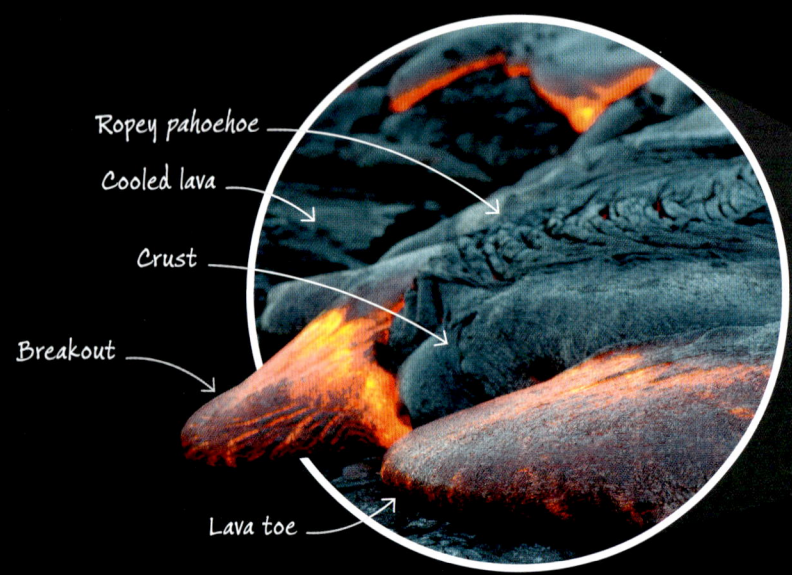

Ropey pahoehoe

Cooled lava

Crust

Breakout

Lava toe

STICKY OR RUNNY?

If lava is very thick and sticky, we describe it as viscous. The viscosity of lava depends on how hot it is and how much silica (silicon dioxide) it contains – the higher the silica content, the thicker the lava. Very viscous lava can't flow easily, and it can cause explosive eruptions. Runny lava is less likely to cause explosions, but it can flow a long way.

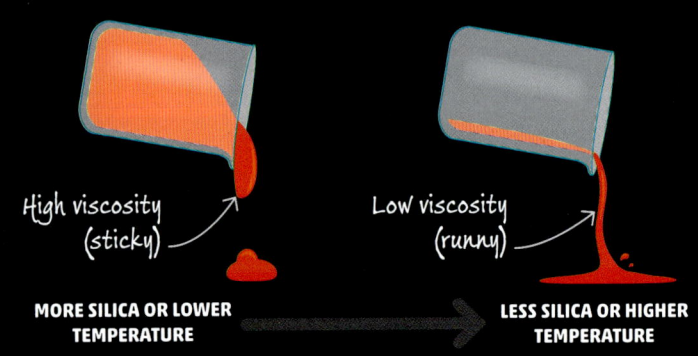

High viscosity (sticky)

Low viscosity (runny)

MORE SILICA OR LOWER TEMPERATURE

LESS SILICA OR HIGHER TEMPERATURE

TYPES OF LAVA FLOW

Lava takes a surprising variety of different forms, depending on its chemical make-up, its temperature, and how much water, gas, and rock crystals it contains. The names *aa* and *pahoehoe* come from Hawaii, where lava flows are common.

AA

Aa (pronounced ah ah) is a thick, rubbly basalt lava that bulldozes anything in its way. The outer edges of the stream harden as they cool, building up to form walls and raising the flow high above ground. The jagged surface of *aa* makes it hard to walk on once it's cooled.

▼ PAHOEHOE

Pahoehoe (pronounced pa-ho-ee-ho-ee) lava is fluid, with a consistency like pancake batter. The front of the flow moves in lobe shapes called toes. It forms beautiful shapes as the surface cools, hardens, and then stretches and folds while the runny interior keeps flowing.

PILLOW LAVA
When lava erupts underwater, it cools quickly and forms pillow shapes. While pillow lava might look unusual, it is actually the most common type of lava, as most lava flows happen in the sea.

BLOCKY LAVA
Blocky lavas are very viscous, thick, and slow-moving compared to other lavas. Sharp-edged blocks form on the surface as the lava cools. These can grow very large and sometimes tumble off the front of the flow.

CARBONATITE LAVA
Carbonatite is a rare, black kind of lava that can only be seen at one volcano in Tanzania in Africa. It contains less silica than other lavas, so it is much runnier. It also erupts at lower temperatures, which is why it doesn't glow.

LAVA TUBES
Lava tubes form in *pahoehoe* as the solidified crust insulates the lava underneath. This allows the lava to flow long distances in underground rivers, which form tube-shaped caves after the lava stops flowing and drains out.

LIQUID ROCK

At temperatures over 700°C (1,300°F), rock melts and becomes lava. *Pahoehoe* lava gets its name from the Hawaiian word for paddling because the ripples in its surface look like the swirling patterns made by an oar in the sea. The lava's cooler outer surface forms an elastic skin that stretches and folds as the runnier interior drags it downhill.

HOW **LAVA** COOLS

Different types of volcanic eruption result in many different kinds of lava, which cool and solidify into a variety of wonderful rock structures called pyroclasts. By studying pyroclasts, volcanologists can work out what kind of eruption a volcano is likely to have in the future.

SCORIA
Scoria is a vesicular volcanic rock – a rock full of vesicles (bubbles) formed by gas inside lava. It has larger and fewer air pockets than pumice, but is denser and doesn't float. The lava that forms scoria is usually less viscous than lava that forms pumice, allowing trapped air to escape more easily.

Pumice is less dense than water

PELE'S HAIR
When lava drips off a cliff or gets shot high in the air, droplets can stretch into long, thin threads. These are called Pele's hair, after the Hawaiian goddess of fire, volcanoes, and creation.

STRANDS OF PELE'S HAIR

Reticulite has so many bubbles that you can often see straight through it.

VOLCANIC GLASS
Obsidian, also known as volcanic glass, is a smooth, dark rock that forms when lava cools too quickly for crystals to form. Just like ordinary glass, it's rich in silicon dioxide (silica). Obsidian breaks into razor-sharp fragments and has been used since the Stone Age to make arrowheads and knives.

FLOATING ROCKS
Explosive volcanic eruptions eject ash and pumice into the air. The sticky magma inside an explosive volcano traps gas, making it foamy and full of tiny gas bubbles (vesicles). It cools to form pumice, a rock that is so light it can float on water. Rafts of pumice can appear floating on the ocean after underwater eruptions.

Sharp, glasslike edges

Thin, broken ends may have detached from a strand of Pele's hair.

PELE'S TEARS
Small blobs of liquid lava form teardrop shapes as they fall to the ground. These are called Pele's tears, and they can form at the end of Pele's hair strands.

CROPS COVERED IN VOLCANIC ASH

VOLCANIC ASH
Ash clouds from volcanoes consist of billions of tiny fragments of volcanic glass. During an explosive eruption, gases in magma expand and shatter the magma into minute flecks that harden in air to become ash. Near Mount Etna in Italy, drivers must replace the tyres on their cars as often as twice a year because volcanic ash on the ground shreds and wears away the rubber.

LAVA BOMBS
These volcanic hazards are often thrown into the air during eruptions. Breadcrust bombs cool and solidify on the outside first, cracking later like breadcrust as gas escapes. Cowpat bombs are still soft as they land and so form flatter, uneven discs.

BREADCRUST LAVA BOMB

COWPAT LAVA BOMB

RETICULITE
This volcanic rock forms in tall lava fountains. It is full of air bubbles and so lightweight that the wind blows it across the ground like tumbleweed.

LAVA TREES
These stumps are made of solidified lava. Flowing lava can engulf living trees, then cool and form a solid crust around them. This leaves a cast of the tree where the real tree – long since burned away – used to be.

LAVA TREES, HAWAII

HOW PYROCLASTIC FLOWS WORK

The deadliest thing a volcano can produce is a pyroclastic flow, which is a dense, ground-hugging avalanche of superhot gas, ash, and rocks that races downhill. Pyroclastic flows demolish and burn everything in their path and can bury the landscape under tons of volcanic debris.

Temperatures can reach 200–700°C (390–1300°F) inside a pyroclastic flow.

▶ MOUNT PINATUBO

The eruption of Mount Pinatubo in the Philippines on 15 June 1991 was the second-largest eruption in the 20th century. It sent pyroclastic flows sweeping down the slopes of the volcano, filling valleys with volcanic deposits. This photo was captured by a photographer who was racing away in the back of a truck.

Dense, scalding ash cloud

The rising ash cloud is called a phoenix cloud.

Large boulders are carried at the bottom.

Flows downhill at great speed

INSIDE A PYROCLASTIC FLOW

Pyroclastic flows have a violent and turbulent interior and carry debris ranging from scalding dust and ash to large boulders. The fast-flowing clouds of rock fragments strip vegetation from the volcano's slope, erode the ground, and set fire to anything in their path.

The pyroclastic flows from Pinatubo filled valleys with volcanic debris up to 200 m (660 ft) thick.

HOW PYROCLASTIC FLOWS FORM

Pyroclastic flows form in several different ways. Some are mostly ash and gas, while others are full of rubble. All hurtle downhill at terrific speed – some as fast as 700 kph (435 mph).

Sideways blast
The volcano erupts sideways rather than vertically.

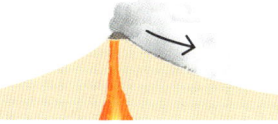

Dome collapse
An eruption shatters an unstable lava dome, creating an avalanche of hot rock.

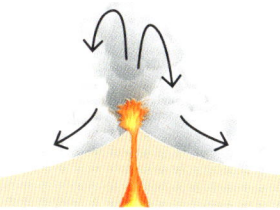

Ash cloud collapse
A large ash cloud partially collapses as heavy material falls back.

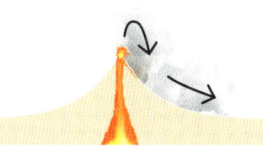

Boiling over
Heavy ash rises a short distance before falling and rolling downhill.

MOUNT VESUVIUS

In 79 CE, the city of Pompeii in Italy was engulfed by pyroclastic flows from Mount Vesuvius. Around 1,500 years later, archaeologists discovered the buried town and found body-shaped cavities. They created casts of the victims, revealing they had died from the scalding heat and from breathing in ash.

KRAKATAU

The 1883 eruption of Krakatau in Indonesia was one of the deadliest in history, killing over 36,000 people. Pyroclastic flows ploughed into the sea, generating tsunamis. In 1927, Anak Krakatau began to emerge from the caldera that was formed by Krakatau, and continues to erupt today.

ANAK KRAKATAU, INDONESIA

HOW A CALDERA FORMS

Massive eruptions can make a volcano collapse inwards and form a giant crater – a caldera. Calderas often fill with water to create lakes or lagoons and are among the most tranquil landscapes on Earth. But these beautiful places are evidence of incredibly violent events that took place in the past.

RISE AND FALL
The city of Pozzuoli in southern Italy sits inside a caldera. Roman ruins that are above sea level today have holes made by marine mussels, which shows that the land has sunk below sea level since Roman times only to rise again. This is evidence of an active magma chamber making the ground rise and fall over time.

▼ CRATER LAKE
Crater Lake in Oregon, USA, is a caldera that formed 7,700 years ago as a result of an explosive eruption. Later eruptions created Wizard Island, a small conical volcano in the lake. At 594 m (1,949 ft) deep, Crater Lake is the USA's deepest lake and the ninth deepest lake in the world. No rivers flow into or out of it – the crystal clear water comes solely from rain and melted snow.

Caldera wall

Crater Lake is 10 km (6 miles) across at its widest point.

SANTORINI

The island of Santorini in Greece is a submerged caldera. Around 3,600 years ago, one of the largest volcanic eruptions in recorded history destroyed the prehistoric city of Akrotiri and triggered tsunamis that devastated nearby islands. The eruption obliterated the centre of the volcano, leaving a ring of islands.

SLOW COLLAPSE

Not all calderas form suddenly. In 2014–15, volcanologists observed the gradual formation of a caldera at Bardarbunga in Iceland. The crater developed over six months as lava slowly drained from the magma chamber to the surface, forming what is now a flat plain.

FORMATION OF CRATER LAKE

A caldera forms when a large eruption empties or partially empties a magma chamber and the volcano collapses into the empty space. Crater Lake formed when a stratovolcano called Mount Mazama self-destructed.

1 PRESSURE BUILDS
Before the eruption, Mount Mazama's peak was 3,650 m (12,000 ft) high, with a vast magma chamber below.

2 ERUPTION
An explosive eruption partly emptied the magma chamber, making its roof unstable.

3 COLLAPSE
The peak collapsed, creating a crater. More than 2,400 m (8,000 ft) of Mount Mazama's height was destroyed.

4 FLOOD
The crater flooded to form a lake. Later eruptions produced lava domes and small cones. One of these became Wizard Island.

Wizard Island

Wizard Island

HOW **HOT SPOTS** WORK

Some of the largest volcanoes on Earth occur in sites that geologists call hot spots. These sites in Earth's crust lie over mantle plumes – columns of hot rock that rise from deep in the mantle or even the core, bringing heat to the planet's surface. Hot-spot volcanoes produce huge volumes of runny, basaltic lava that can build up on the sea floor to form islands. In places like Hawaii, chains of volcanoes have formed as the ocean floor has slowly moved across the hot spot.

▼ KĪLAUEA, HAWAII

Kīlauea on Hawaii is one of the most active volcanoes in the world. At its summit is a crater called Halemaumau that sometimes fills with molten rock to form a lava lake. Lava lakes are rare – only another six exist on Earth. Eruptions from fissures in the sides of Kīlauea can drain this deep lake, producing rivers of lava that flow all the way to the sea.

THE HAWAIIAN ISLANDS

The Hawaiian Islands formed as the Pacific tectonic plate slowly moved over a hot spot. The hot spot melts the lithosphere above, creating magma that erupts to produce one new island after another. Only the newer islands still have active volcanoes. Further away, the older volcanoes have become dormant or extinct, and many have eroded and sunk below the ocean surface.

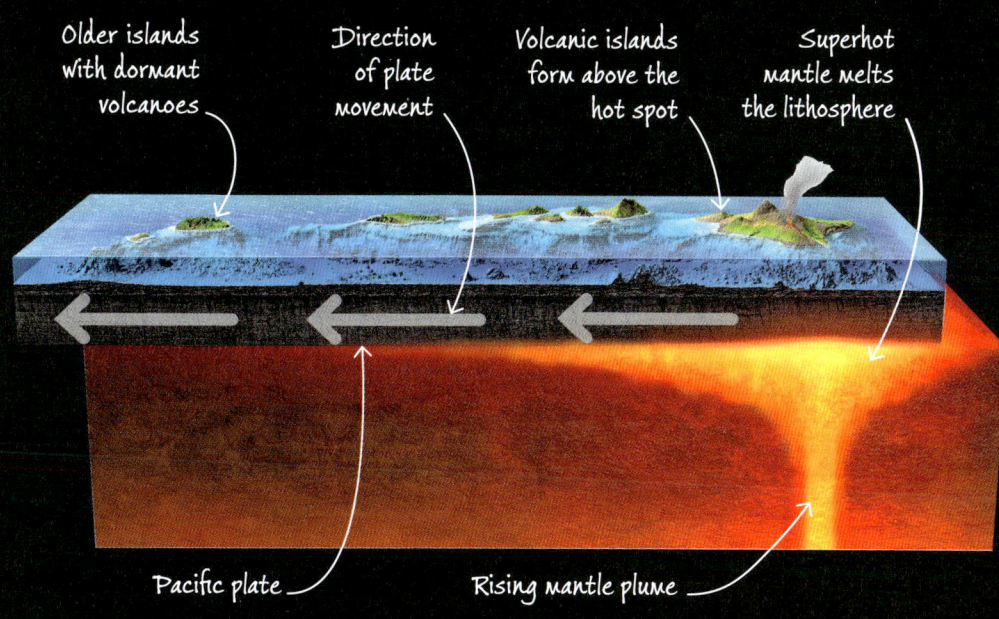

Older islands with dormant volcanoes

Direction of plate movement

Volcanic islands form above the hot spot

Superhot mantle melts the lithosphere

Pacific plate

Rising mantle plume

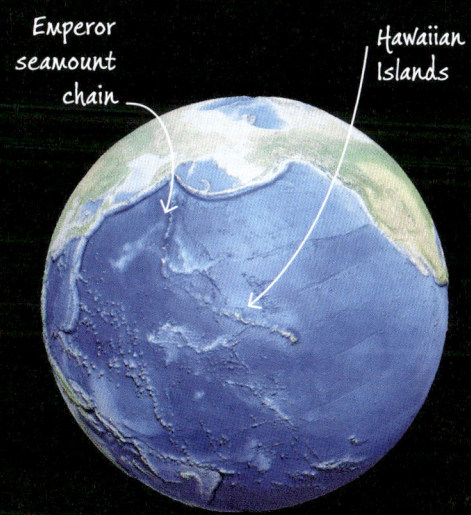

Emperor seamount chain

Hawaiian Islands

HAWAIIAN-EMPEROR SEAMOUNT CHAIN

The Hawaiian Islands are part of a longer chain of islands and underwater mountains (seamounts) extending for 6,200 km (3,900 miles): the Hawaiian-Emperor seamount chain. This vast structure formed over 85 million years as the Pacific plate moved across a hot spot. The shape of the chain tells us about the plate's movement, with a bend marking a change in direction.

GODDESS OF VOLCANOES

According to Hawaiian mythology, the Halemaumau crater on Kīlauea is home to Pele, goddess of volcanoes and fire. Pele had a fiery personality and fought with her sister, the goddess of the sea. Pele was driven from island to island, each time digging a fire pit for herself and creating new eruptions. These ancient stories explained how the Hawaiian volcanoes formed, long before scientific theories developed.

HOW ATOLLS FORM

If you take a flight over a tropical ocean, you might spot ring-shaped islands or ring-shaped reefs with blue lagoons in the middle. These are atolls and are built by small sea creatures called corals. Corals grow in large colonies inside hard skeletons made of calcium carbonate. Over thousands of years, the skeletons of dead corals build up to form coral reefs, which are home to an amazing diversity of life.

▼ BORA BORA

The Bora Bora atoll formed around an extinct volcano in the South Pacific. The island is encircled by a lagoon, which is surrounded by a barrier reef. The barrier reef protects the lagoon from the ocean waves, making the calm waters a haven for stingrays, barracudas, sharks, and other fish.

Extinct volcanoes no longer have magma in their magma chamber and so will not erupt.

Lagoons are shallow, bodies of water protected from the sea by land or reefs.

SINKING ISLANDS

Almost 200 years ago the British naturalist Charles Darwin produced a map of all the atolls he found during a round-the-world sea trip, and proposed the theory that atolls form around sunken islands.

❶ ISLAND FORMATION
A hot-spot volcano forms an island, and a fringing reef then grows around the island. Once the magma chamber cools and solidifies, the volcano is extinct.

Volcanic island

Fringing reef

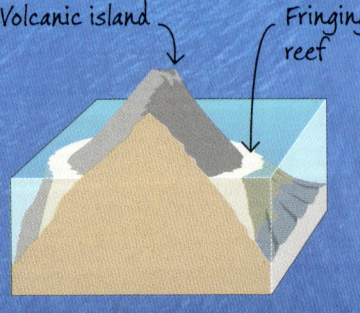

❷ EROSION AND SINKING
The island sinks due to erosion, sinking of the sea floor, or both. The reef grows upwards, staying just below sea level. A lagoon forms between the land and the reef.

Lagoon

Reef grows upwards

Island sinks

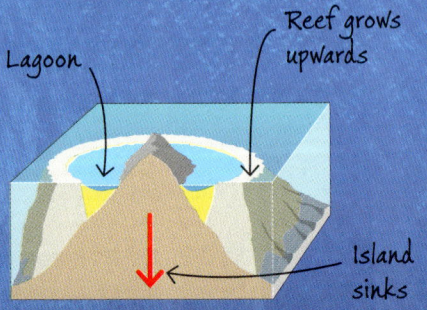

CORAL COLONIES

Coral colonies come in all sorts of colours, shapes, and sizes. Some look like the branches of a tree, and some look like leaves or flowers. Fan-shaped corals reach into the water, trying to trap food, while vase-shaped corals are a hiding place for fish.

AN ATOLL NATION

The Maldives is a country in the Indian Ocean made up of around 1,200 small coral islands arranged in rings to form 26 huge atolls. The Maldives is the lowest-lying country in the world, with the highest point being only 2.4 m (8 ft) above sea level.

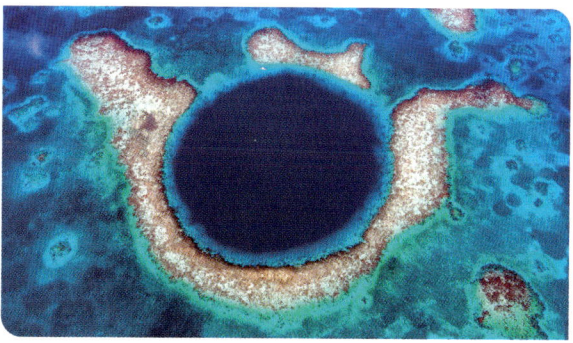

GREAT BLUE HOLE

Not all atolls develop from sinking islands. The Great Blue Hole in the Caribbean Sea was once a cave on dry land, when sea levels were lower than today. The cave's roof collapsed, forming a sinkhole, and the sea level later rose and filled it, creating an atoll.

As the coral on the lagoon side dies, it breaks down to form sand.

Waves break on the outer edge of the reef.

3 ATOLL GROWS
Eventually the island disappears, leaving just a circular reef – an atoll. The outside of the atoll grows upwards and outwards, and the inside gets broken up, depositing sand on the bottom of the lagoon.

Inside of reef is eroded

Reef grows upwards and outwards

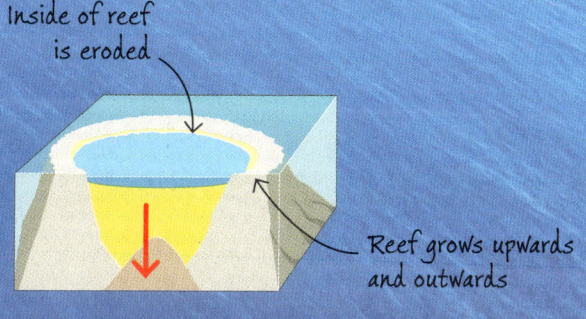

NUKUORO ATOLL

Nukuoro in the Pacific is another atoll that formed due to changing sea levels. Long ago, a fall in sea level turned the shallow limestone sea floor here into land. The limestone eroded to form a bowl shape, which became an atoll when the water rose again.

HOW GEYSERS AND MUD POTS WORK

The ground beneath your feet contains more than just rock. It also contains groundwater, which trickles though a maze of hidden channels and soaks into layers of softer rock like water into a sponge. When volcanoes heat this groundwater, amazing geothermal features can form. Hot water and steam gush through cracks and other spaces to the surface, where the hot, mineral-rich water forms hot springs, geysers, mud pots, and unusual rock formations.

▶ STROKKUR GEYSER, ICELAND

Geysers are volcanic springs that shoot out boiling water and steam. The name comes from the Icelandic word *geysir*, meaning "to gush". Strokkur, a fountain geyser, is Iceland's most regularly erupting geyser. Every 6–10 minutes, a fountain of scalding water shoots 20 m (65 ft) high, though it has been known to reach twice this height.

HOW GEYSERS WORK

Geysers form when heat from volcanic activity boils groundwater but the steam gets trapped, builds up, and periodically erupts. At Strokkur, the trapped pocket of steam expands and pushes up the pool before bursting out. The sudden release of pressure creates an explosive fountain of hot water and steam. The pressure then starts to build again.

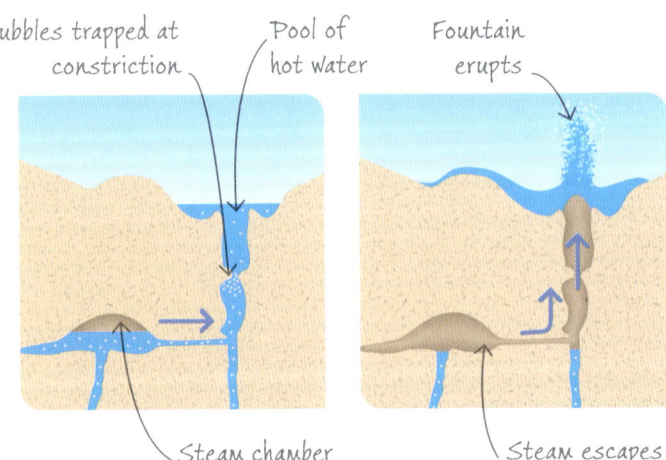

Bubbles trapped at constriction

Pool of hot water

Fountain erupts

Steam chamber

Steam escapes

❶ STEAM BUILDS UP
Bubbles of steam rise to a chamber that traps them. A constriction in the outlet traps more bubbles, causing pressure below to build up.

❷ ERUPTION
Trapped steam escapes, releasing pressure. The sudden fall in pressure allows the scalding water to boil, creating an explosive fountain.

LIFE IN HOT SPRINGS
Not all hot springs are boiling – some are cool enough to bathe in. Even the hottest springs have life. Microorganisms called extremophiles can withstand the heat and can obtain energy from minerals dissolved in the water. Some scientists think the earliest forms of life on Earth may have lived this way.

CONE GEYSERS
Whereas fountain geysers erupt from pools, cone geysers erupt from chimney-like mounds of sinter, a rocky mineral deposited by the water. The Fly Ranch geyser in Nevada, USA, owes its vibrant colours to thermophilic (heat-loving) algae.

MUD POTS
Mud pots are pools of boiling mud. They form where a hot spring has relatively little water, and the underlying rock is eaten away by acidic gases and by extremophiles. This results in a gooey, grey mud through which steam and hot water erupt. Some mud pots are cool enough to sit in. The mud is thought to be good for the skin.

LIMESTONE POOLS

The water that bubbles out of hot springs is rich in minerals dissolved underground. As the water trickles away and evaporates, the minerals crystallize on the ground as a hard crust. In the mountains of Turkey, this process has created a natural wonder – a terrace of white limestone pools brimming with turquoise water. Called Pamukkale ("cotton castle" in Turkish), it has attracted tourists for more than 2,000 years.

HOW
SUPERVOLCANOES
WORK

The most powerful volcanoes can produce catastrophic supereruptions. A supereruption can bury a whole country under a blanket of ash and release enough volcanic gas to change the climate. At the eruption site, pyroclastic flows destroy everything in their path, and the ground collapses into a gigantic crater (a caldera). Thankfully, supereruptions are rare – the last one was 27,000 years ago.

▲ YELLOWSTONE
The Yellowstone volcano in the USA has had three supereruptions in the last 2 million years. Each lasted for decades and covered North America in ash. Today, the caldera is a popular beauty spot, thanks to its spectacular hot springs (above) and geysers, kept active by the giant magma chamber still lurking deep below.

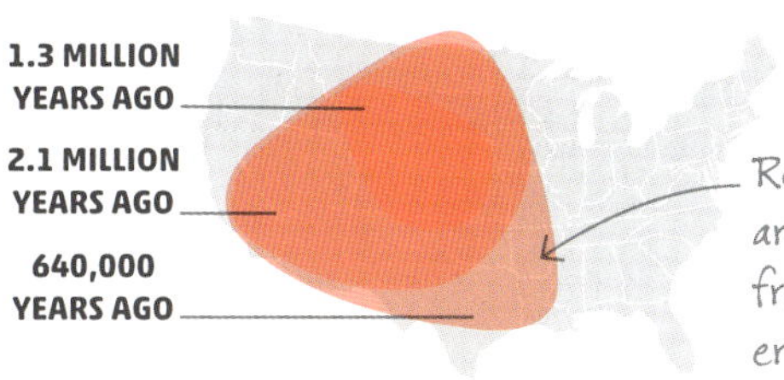

1.3 MILLION
YEARS AGO

2.1 MILLION
YEARS AGO

640,000
YEARS AGO

Red zones
are ash falls
from past
eruptions.

PAST ERUPTIONS

Traces of ash from past eruptions at Yellowstone show how devastating a supereruption can be. Most of Yellowstone's eruptions are now small lava flows, so the next eruption is unlikely to be a supereruption.

These rock layers, formed in lava floods, stretch thousands of miles across India.

LAVA FLOODS

Some of the mass extinctions in Earth's past may have been caused by supereruptions. About 65 million years ago, just before the dinosaurs died out, massive floods of lava spewed from volcanoes in India for thousands of years. The lava covered much of western India in igneous rock, forming geological structures now known as the Deccan Traps.

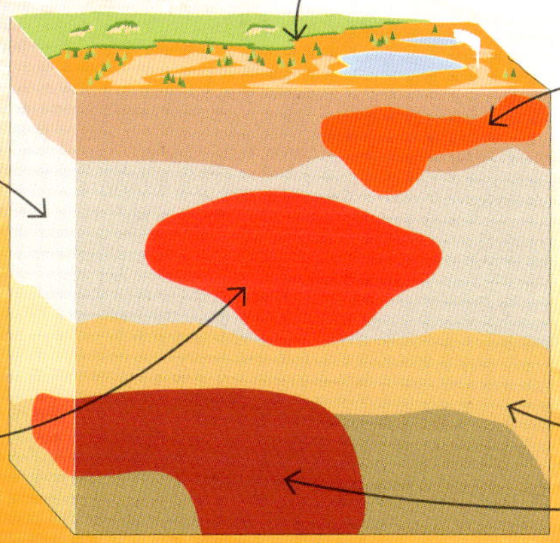

VOLCANIC WINTERS

In 1815, the eruption of Mount Tambora in Indonesia led to the "Year Without Summer". Gases released by the volcano interacted with water vapour in the atmosphere and blocked out the Sun. This caused global failed harvests and famines. Many horses had to be killed due to lack of food, prompting the invention of the bicycle.

Grand Prismatic Spring is a hot spring heated by the Yellowstone volcano.

Yellowstone National Park

Upper magma chamber containing thick rhyolite magma, which can cause explosive eruptions. Only 5–15% of the chamber is molten.

Crust

Lower magma chamber containing runny, basaltic lava. Only 2% of the chamber is molten.

Upper mantle

Mantle plume

UNDER YELLOWSTONE

Yellowstone's volcano sits over a mantle plume – a column of hot magma rising from deep in Earth's mantle. Heat from the plume has melted parts of the crust, creating two giant magma chambers. These are not caverns full of liquid rock, but hot zones containing scattered pockets of liquid magma. When enough pockets join, an eruption can occur.

HOW EARTHQUAKES WORK

Earthquakes happen when huge areas of rock grind and slip past each other at fractures called faults. Thousands of tiny earthquakes happen every day, but the most violent ones happen at the biggest faults, which are near the boundaries between tectonic plates. When these strike in highly populated areas, they can be devastating.

▶ ALASKA, 2018
When movement along faults is sudden, energy is released as powerful seismic waves. These cause the ground to shake, buckle, and rupture, which can demolish roads and buildings in seconds.

LOCK AND SNAP
Just as a squeezed spring stores energy, the rock in Earth's crust stores energy when compressed or stretched. If this stored energy is released by a sudden movement, it causes an earthquake. The hypocentre of an earthquake is where the movement happens. The epicentre is the point on the surface directly above.

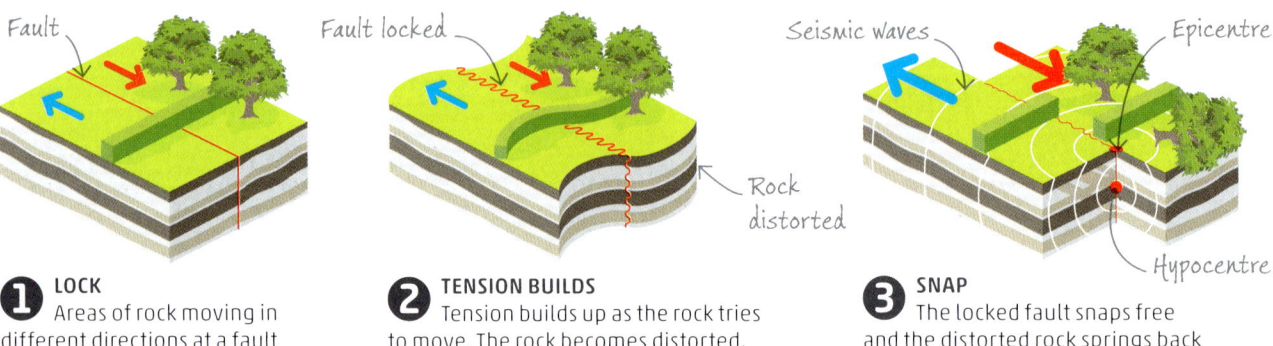

❶ LOCK
Areas of rock moving in different directions at a fault snag together and get stuck.

❷ TENSION BUILDS
Tension builds up as the rock tries to move. The rock becomes distorted, storing energy like a spring.

❸ SNAP
The locked fault snaps free and the distorted rock springs back into shape suddenly. The release of energy causes an earthquake.

Fault · *Fault locked* · *Rock distorted* · *Seismic waves* · *Epicentre* · *Hypocentre*

EARTHQUAKE SCALES
Earthquakes are measured in two ways. The magnitude scale (below) is based on the energy released, which is measured with vibration-detecting instruments called seismometers. The intensity scale is based on how much damage is done.

Strong enough to feel, but not strong enough to cause much damage

Chance of damage near the epicentre

Strong vibrations that may cause a lot of damage near the epicentre

Major earthquake that is likely to cause extensive damage over a large area

Huge earthquake that is likely to completely destroy buildings around its epicentre

❷ VERY MINOR   **❸ MINOR**   **❹ LIGHT**   **❺ MODERATE**   **❻ STRONG**   **❼ MAJOR**   **❽ DEVASTATING**

MAGNITUDE SCALE

LIQUEFACTION

If the ground is wet and made of soil or loose material, a process called liquefaction can occur during an earthquake. The shaking moves the loose ground so much that it flows like a liquid and engulfs objects. Cars and buildings sink into the ground, while buried pipes and cables float to the surface.

EARTHQUAKE PROTECTION

Earthquakes are hard to predict, but scientists can detect warning signs, including foreshocks and changes in ground height. Most deaths are caused by the collapse of buildings, so an effective form of protection is to design buildings that can withstand being shaken, with shock-absorbing foundations and reinforced steel frames.

SAN FRANCISCO EARTHQUAKE

The deadliest earthquake in US history took place in San Francisco in 1906. It was caused by the San Andreas fault between the Pacific and North American tectonic plates. The Pacific plate lurched 10 m (33 ft) north, causing an earthquake with a magnitude of 7.9. The shaking and subsequent fires destroyed 80 per cent of the city. It was one of the first earthquakes to be recorded on film.

MOTION OF MATERIAL

WAVE DIRECTION

Pushing a stretched spring forwards mimics the compression and stretching of P-waves.

HOW SEISMIC WAVES WORK

Earthquakes unleash vast amounts of energy, triggering powerful vibrations that spread through the ground at thousands of miles an hour. These vibrations are called seismic waves. There are several different types, with some travelling only through rock on Earth's surface and others passing right through our planet's core. By studying these waves, scientists can figure out not only why earthquakes happen but also how Earth's inner layers work.

▲▼ P- AND S-WAVES

Surface waves travel in the rock layers just below Earth's surface, and body waves travel through the inside of Earth. Body waves are either primary (P-) waves or secondary (S-) waves. P-waves travel by compressing and expanding the rock they pass through, and S-waves travel in a side-to-side motion. P-waves and S-waves travel on a curved path through Earth because of changes in density in the mantle and the core.

Flicking a stretched spring sideways mimics the movement of S-waves.

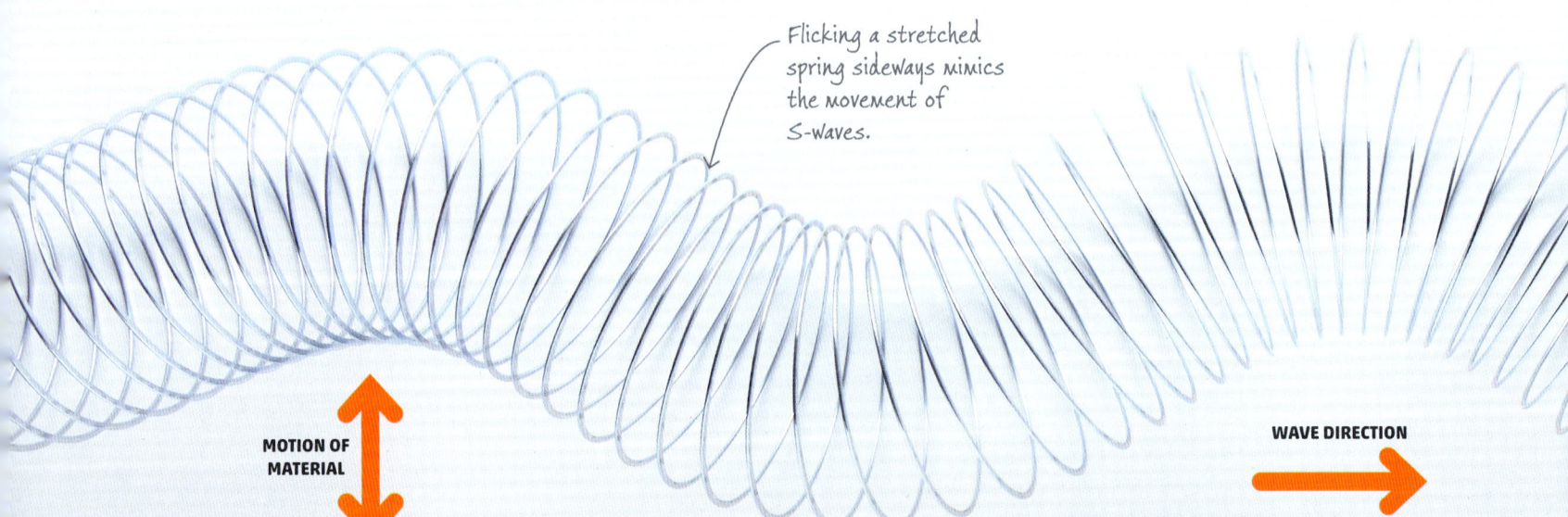

MOTION OF MATERIAL

WAVE DIRECTION

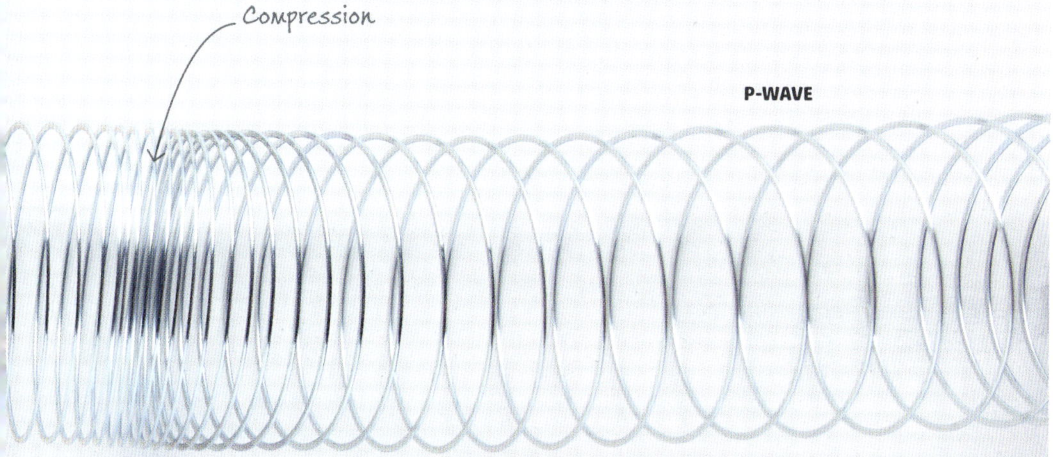

Compression

P-WAVE

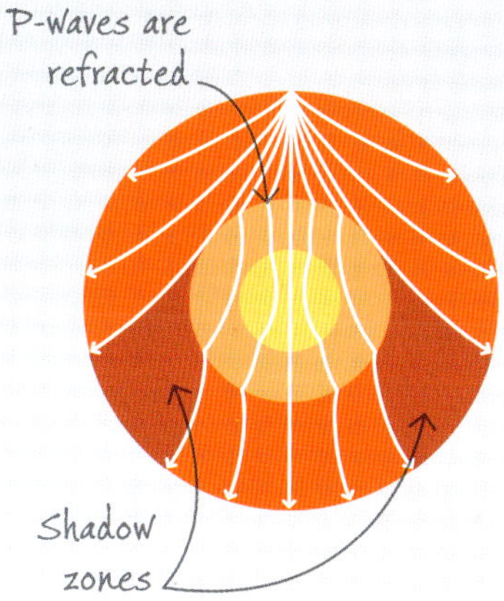

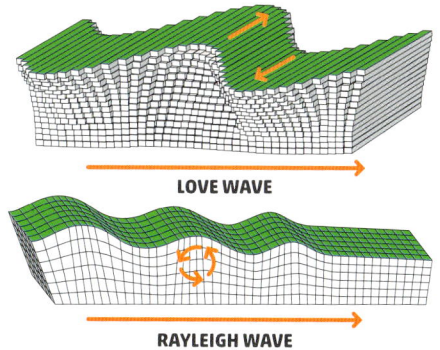

LOVE WAVE

RAYLEIGH WAVE

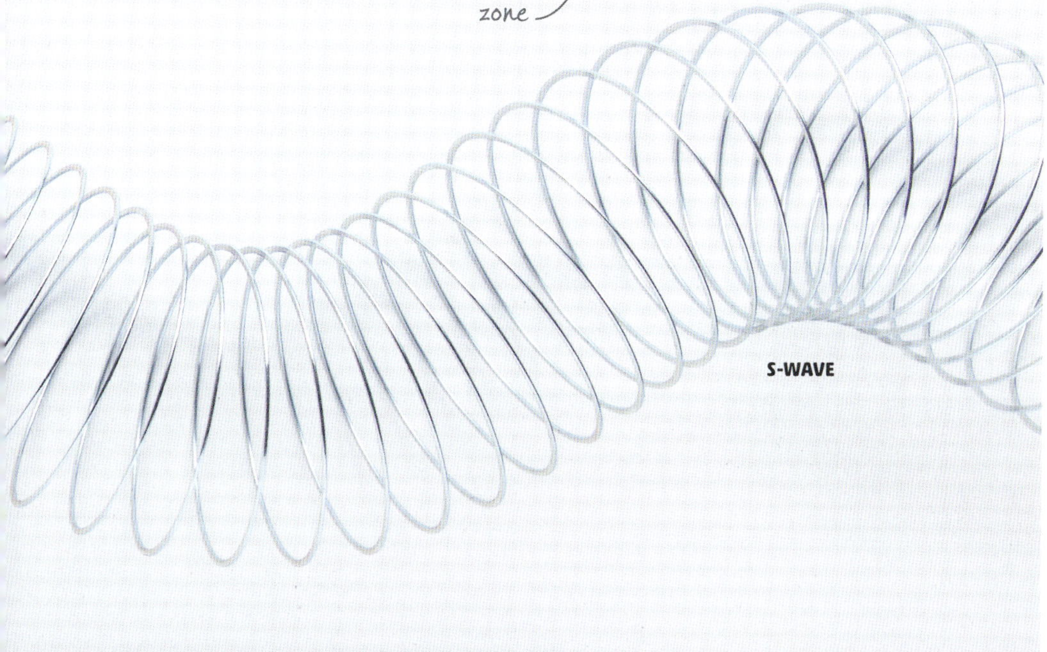

SURFACE WAVES

There are two types of surface wave. Love waves move the ground from side to side, a bit like S-waves trapped at the surface. These cause the most damage to roads and buildings. Rayleigh waves move the surface in a circular motion, a bit like ocean waves.

P-waves are refracted

SEEING INSIDE EARTH

When an earthquake happens, seismic waves travel through Earth and can be detected around the globe. However, there are "shadow zones" where waves are not recorded. By studying the locations of shadow zones from different earthquakes, geologists worked out the size of Earth's inner layers and whether they were solid or liquid.

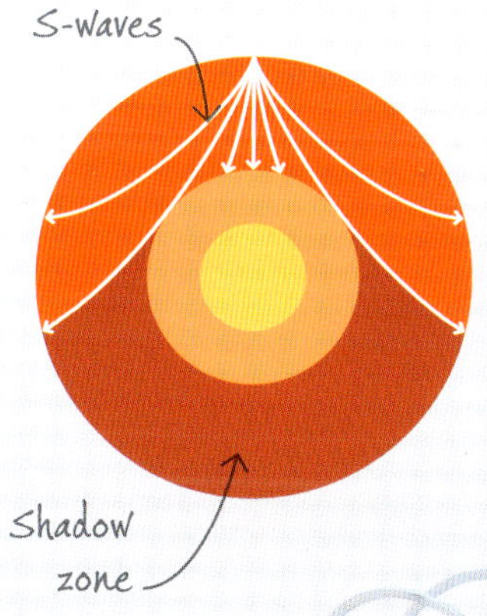

Shadow zones

P-WAVES

P-waves can travel through solids and liquids, but they are refracted (bent) when they pass from one to the other. Their shadow zones reveal the size of Earth's core and the presence of a solid inner core.

S-waves

S-WAVES

S-waves can travel through solids but not liquids. The large shadow zones on the far side of Earth show that the planet has a molten outer core.

Shadow zone

The horizontal lines show the strength of the seismic waves.

SEISMOMETERS

Scientists use instruments called seismometers to measure seismic waves and study earthquakes. The seismometer is securely attached to the ground so that it shakes when an earthquake happens. The motion was traditionally recorded by a pen drawing on a drum of paper, but modern seismometers record their data digitally.

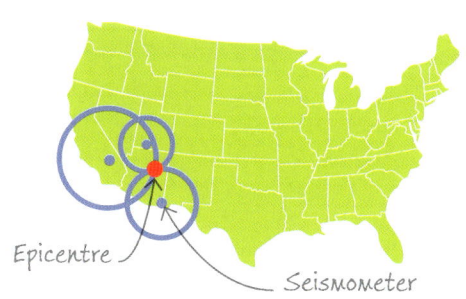

Epicentre

Seismometer

FINDING THE EPICENTRE

To work out exactly where the epicentre of an earthquake is, seismologists need three seismometers in different places. Each one can calculate the distance to the epicentre by measuring when P- and S-waves arrive (P-waves are faster). When the distance from all three seismometers is known, drawing intersecting circles on a map pinpoints the epicentre.

S-WAVE

HOW TSUNAMIS WORK

When an earthquake, volcanic eruption, or landslide happens at sea, it can cause a powerful wave called a tsunami. Tsunamis are usually no taller than ordinary waves but they are thousands of times longer and can race across the open ocean at speeds of 800 kph (500 mph). Larger tsunamis build to a great height as they reach the coast, causing catastrophic floods as they rush inland.

TSUNAMI FORMATION

Most tsunamis are created by earthquakes that cause sudden movements of the sea floor. This displaces a colossal volume of ocean water, producing tsunami waves that radiate outwards.

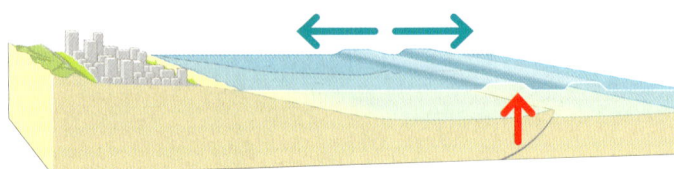

1 SEA FLOOR DISPLACEMENT
The sea floor lurches upwards, creating a wave that spreads in all directions. The wave is low but hundreds of miles long and moves at terrific speed.

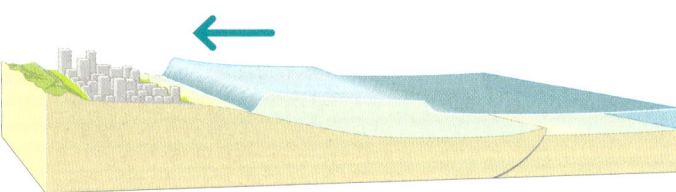

2 WAVES BUILD
When a wave reaches the shore, its front slows down but the rear keeps moving. As a result, water accumulates and the wave grows taller. Water may retreat from the coast before the tsunami hits, as the trough (low point) of the wave may arrive first.

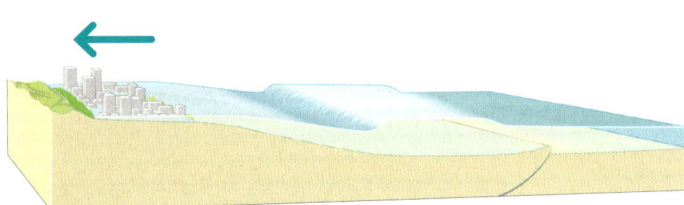

3 COASTAL FLOODING
Now higher than the land, the tsunami washes inland faster than a person can run. The surge of water can carry boats, cars, trees, and debris from demolished buildings.

▼ TŌHOKU TSUNAMI

In 2011, Japan was shaken by the largest earthquake in its history, caused by a sudden slip between the Eurasian and Pacific tectonic plates deep under the Pacific Ocean. The whole of Japan's main island lurched eastwards by 2.4 m (8 ft), and the sea floor sprang upwards by 7 m (23 ft), triggering a tsunami that struck Japan ten minutes later. The wave reached 40.5 m (133 ft) at its highest point and surged inland for up to 10 km (6 miles). The image below – a still from a bystander's video – shows the wave breaching tsunami defences at Miyako.

TSUNAMI WARNINGS

Tsunamis can't be predicted, but tsunami buoys and sensors on the sea floor can detect them and give an early warning. After an earthquake, a tsunami alert is sent to coastal communities that could be affected. The sea draining away quickly from a beach is a sign that a deadly wave is coming and that people need to move to higher ground.

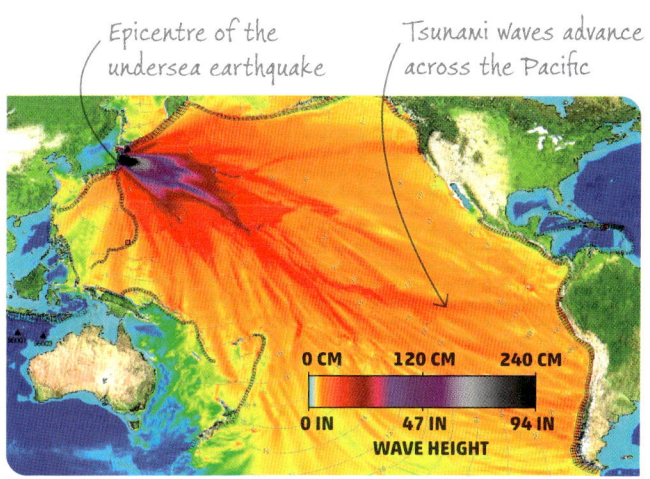

Epicentre of the undersea earthquake

Tsunami waves advance across the Pacific

0 CM	120 CM	240 CM
0 IN	47 IN	94 IN

WAVE HEIGHT

ACROSS THE OCEAN

Tsunami waves travel outwards from their source. They carry a huge quantity of energy and can travel across an entire ocean. The 2011 tsunami that struck Japan also travelled 9,000 km (5,600 miles) across the Pacific Ocean to hit California with 2.7 m (9 ft) waves just a few hours later.

BANDA ACEH BEFORE TSUNAMI

BOXING DAY TSUNAMI

The deadliest tsunami on record took place on 26 December 2004 after a massive earthquake in the Indian Ocean. The tsunami killed around 230,000 people and caused widespread destruction along the coasts of the Indian Ocean.

BANDA ACEH AFTER TSUNAMI

Earth's surface is in a continual state of change. Over millions of years, **tectonic forces** raise **mountains** and reshape **continents**. At the same time, the forces of **weathering** and **erosion** wear away the land, turning solid rock into sand and mud. This endless cycle of creation and destruction has created all the world's **landscapes**, from alpine valleys to desert canyons and coastlines.

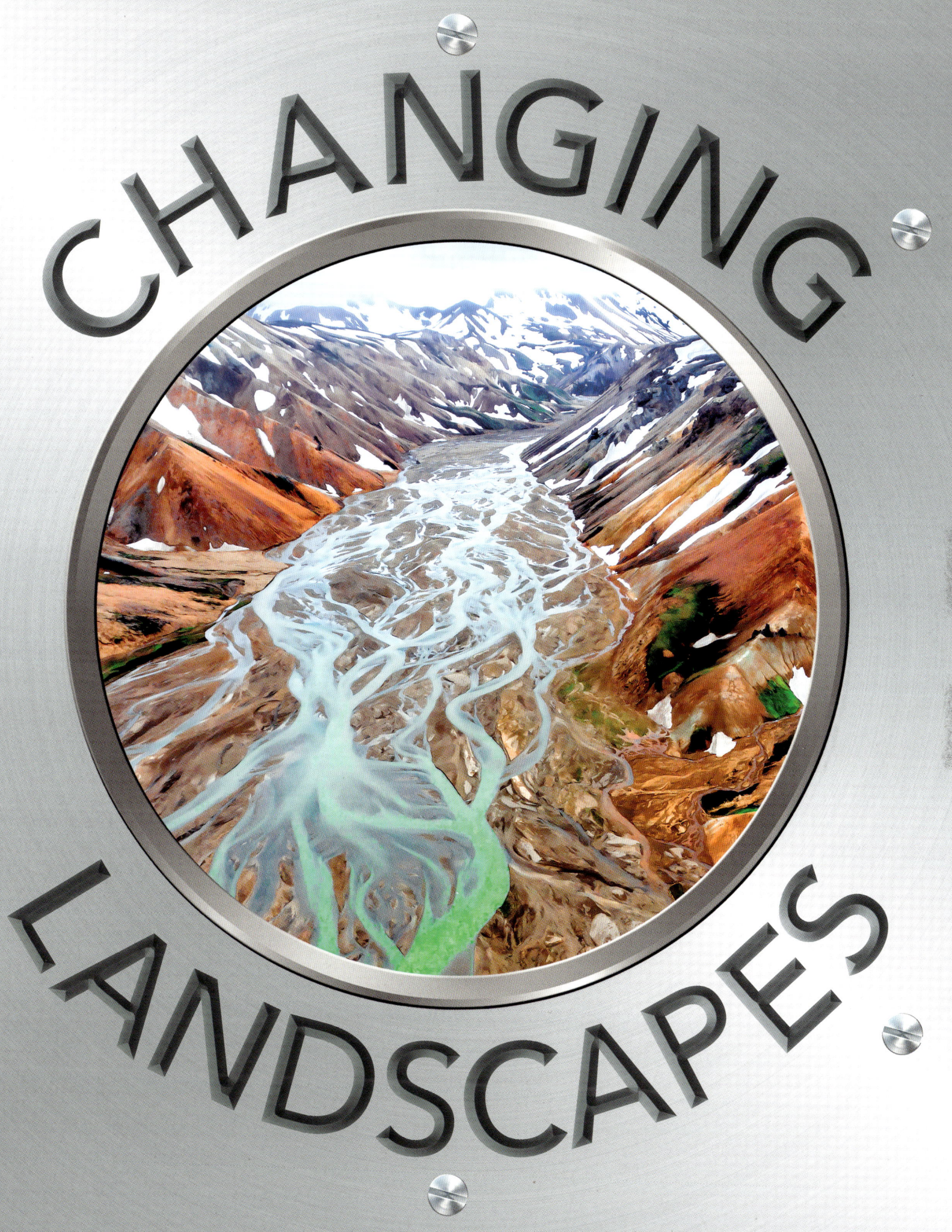

CHANGING · LANDSCAPES ·

HOW LANDSCAPES FORM

Earth's surface is dynamic, which means that it's continually changing. Some changes are so slow that we barely notice them, but something that looks trivial – like rain falling on a hillside – can wear down a whole mountain range given enough time. Other processes, such as volcanic eruptions and landslides, can cause sudden, dramatic changes to landscapes. Every landscape has its own story to tell.

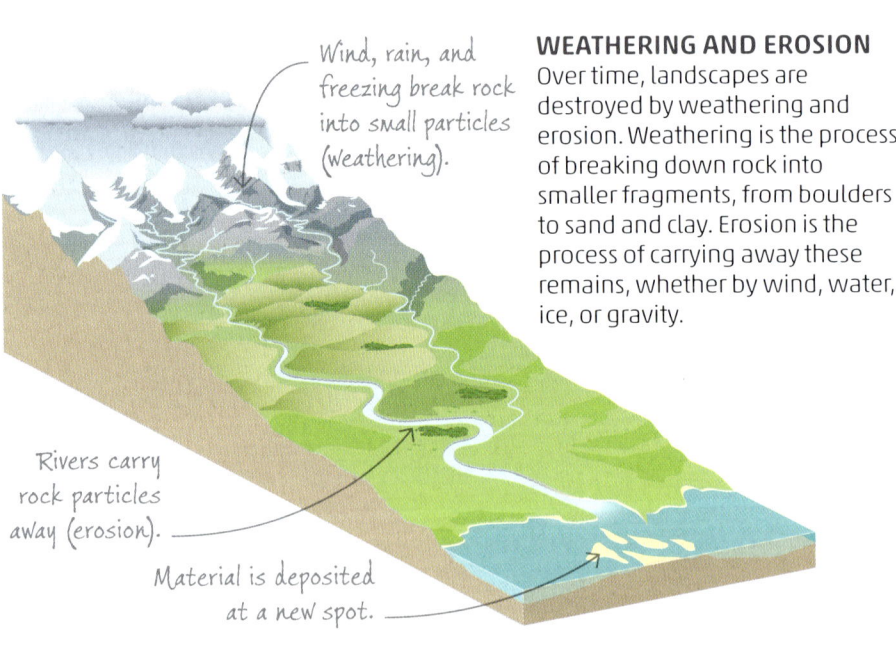

Wind, rain, and freezing break rock into small particles (weathering).

Rivers carry rock particles away (erosion).

Material is deposited at a new spot.

WEATHERING AND EROSION
Over time, landscapes are destroyed by weathering and erosion. Weathering is the process of breaking down rock into smaller fragments, from boulders to sand and clay. Erosion is the process of carrying away these remains, whether by wind, water, ice, or gravity.

KEY TO THE PAST
Mary and Charles Lyell were a geologist couple who visited the Swiss Alps for their honeymoon in 1832. They formed a theory that Earth's landscapes are shaped by gradual processes that have acted for vast spans of time and continue today. As Charles Lyell famously said, "the present is the key to the past".

MARY HORNER LYELL

SIR CHARLES LYELL

▼ ALPINE VALLEY
The Alps mountain range rose bit by bit as tectonic plates carrying Africa and Europe collided, folding and buckling layers of rock that once lay on the sea floor. During the Ice Age, vast glaciers carved deep valleys, like Lauterbrunnen in Switzerland.

Alpine glaciers form where centuries of snow build up on mountain peaks.

❷

❸

❹

❺

Lauterbrunnen valley was filled with a glacier during the Ice Age. Glaciers flow slowly and carve out U-shaped valleys with steep sides and wide, flat bottoms.

Snow and ice break down rock by seeping into crevices and expanding with each cycle of freezing and thawing.

Waterfalls cut through the cliffs creating dramatic gorges.

1 SCREE
Broken rock fragments caused by weathering collect on steep slopes, forming heaps of loose debris called scree.

2 RAPIDS
Rapids are areas of fast-flowing water in shallow rivers or streams with a rocky bottom. They form where the ground is steep and rocky.

3 CAVES
Caves form where groundwater flows through soft rock such as limestone. Limestone reacts with natural acidity in rain and slowly dissolves.

4 FLOODPLAINS
Rivers carry away sediment and deposit it in the valleys, where it forms lush floodplains.

5 MEANDERS
Meanders are bends in rivers. They form when sediment is eroded on the outside of bends and deposited on the inner edges.

❶ BEFORE COMPRESSION
Beds of coloured sand representing sedimentary rock strata are carefully laid down in flat layers. Deeper layers represent older layers of rock.

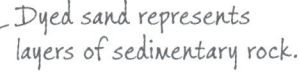

Dyed sand represents layers of sedimentary rock.

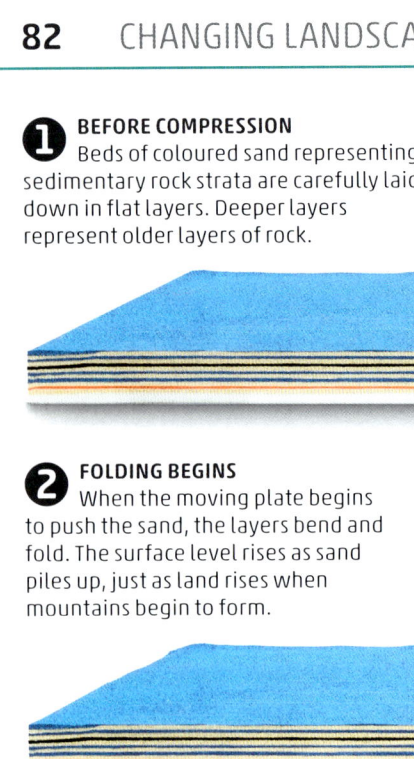

❷ FOLDING BEGINS
When the moving plate begins to push the sand, the layers bend and fold. The surface level rises as sand piles up, just as land rises when mountains begin to form.

Layers begin to fold

❸ FAULTS DEVELOP
Fractures develop when the layers are pushed too much to fold any further. These fractures are known as thrust faults.

Thrust fault

As folding continues, older rock is pushed over younger layers, forming a structure called a nappe.

❹ MOUNTAINS RISE
The crust continues to thicken, raising the sand to form mountains. In the real world, the thickening pushes downwards too, giving mountains very deep roots.

Ridges and valleys form on the surface

▼ MODELLING MOUNTAINS
To understand mountain formation, scientists use sandbox models. Layers of sand representing layers of rock in Earth's crust are slowly compressed by a machine. The sand layers fold and then break along faults, as happens in Earth's crust. The folded layers pile up, making the crust thicker and raising the land.

Far from the zone of folding, the tectonic plate remains thin and undeformed.

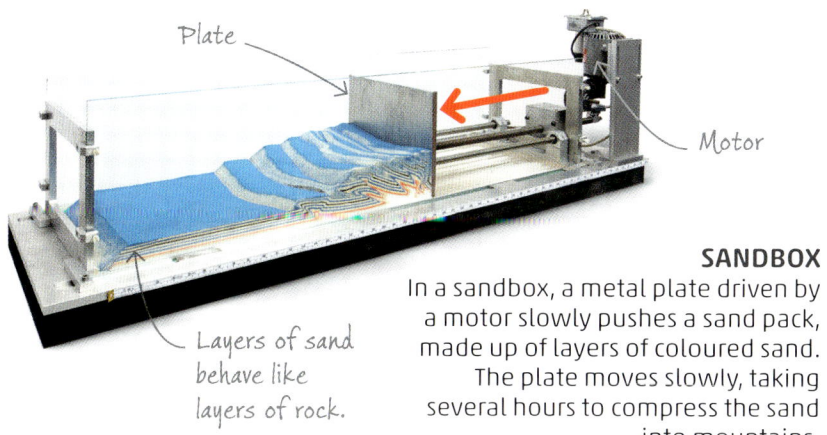

Plate

Motor

Layers of sand behave like layers of rock.

SANDBOX
In a sandbox, a metal plate driven by a motor slowly pushes a sand pack, made up of layers of coloured sand. The plate moves slowly, taking several hours to compress the sand into mountains.

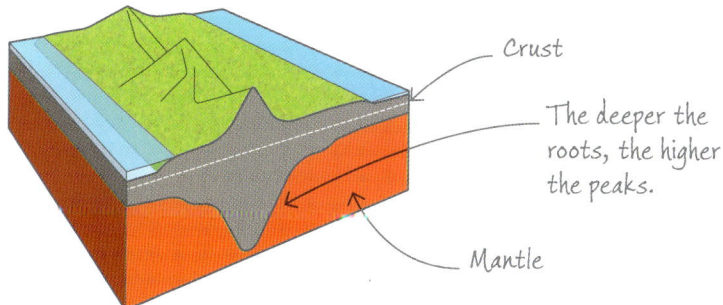

Crust

The deeper the roots, the higher the peaks.

Mantle

DEEP ROOTS
Mountains don't just have high peaks – they have deep roots too. A mountain range "floats" on the soft but denser mantle rock below, its roots submerged. As the peaks are worn away by erosion, the whole mountain range bobs up like an iceberg in water, keeping the peaks high.

FOLDED ROCK
Erosion can make the internal folds of mountains visible. The mountains of Crete in Greece formed when Africa and Europe collided, crumpling layers of sedimentary rock that were once on the sea floor.

FOLDED LIMESTONE, CRETE

HOW
MOUNTAINS
RISE

Mountain ranges form at the boundaries between tectonic plates. The world's tallest mountains – the Himalayas – are still rising today by about 1 cm (0.4 in) a year as the plate carrying India collides with the Eurasian plate. These processes play out over millions of years, but scientists can simulate them in a few hours by using models made of sand.

A fold that bends upwards like an arch is called an anticline.

Foothills

A thrust fault is a sloping fracture that moves older rocks over younger rocks.

A fold that bends downwards in a u-shape or v-shape is called a syncline.

HOW RIFT VALLEYS FORM

The process of stretching and breaking up a continent is called rifting. A rift valley takes millions of years to form and eventually evolves into a new ocean.

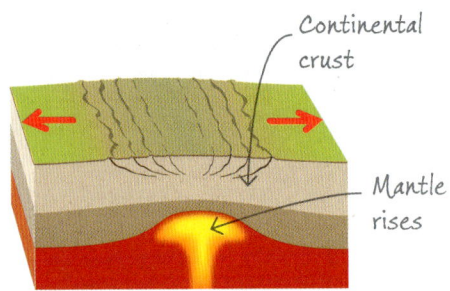

Continental crust

Mantle rises

1. PLATES DIVERGE

When tectonic plates move away from each other, they stretch and thin the crust. The mantle rises and partly melts, forming magma.

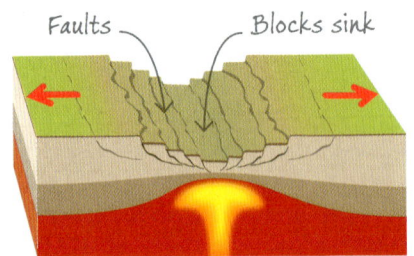

Faults   Blocks sink

2. VALLEY FORMS

The crust breaks along faults, and large blocks sink. This forms an elongated valley called a continental rift.

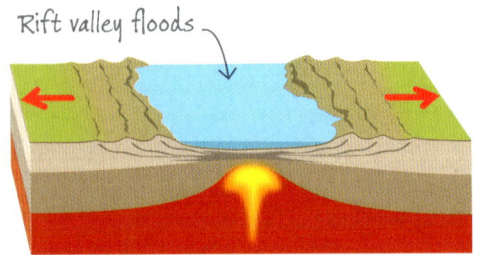

Rift valley floods

3. OCEAN FORMS

As the rifting continues, the land sinks below sea level and floods, forming a linear ocean like the present-day Red Sea.

HOW RIFT VALLEYS SINK

In some parts of the world, neighbouring tectonic plates are pulling apart. When this happens, the crust between them thins and breaks up. Magma wells up from the mantle, oozing through cracks and creating new crust. This process can happen on the sea floor or on land. In the sea, it creates new ocean basins. On land, it tears apart whole continents, forming mountains and rift valleys.

▶ THE GREAT RIFT VALLEY

Africa's Great Rift Valley stretches for more than 3,000 km (1,864 miles) from Jordan to Mozambique, and is slowly tearing the continent apart. The crust here has stretched and broken into blocks separated by deep cracks called faults. Some sections have sunk, forming steep-sided valleys called grabens, flanked by highlands. Magma rising from the mantle feeds the Rift Valley's many volcanoes, including Mount Kilimanjaro – Africa's highest peak.

MODELLING RIFTS

Geologists use sandbox models to study how rift valleys form. Layers of coloured sand on a moving base are slowly pulled apart. A process that can take millions of years can be modelled in a few hours.

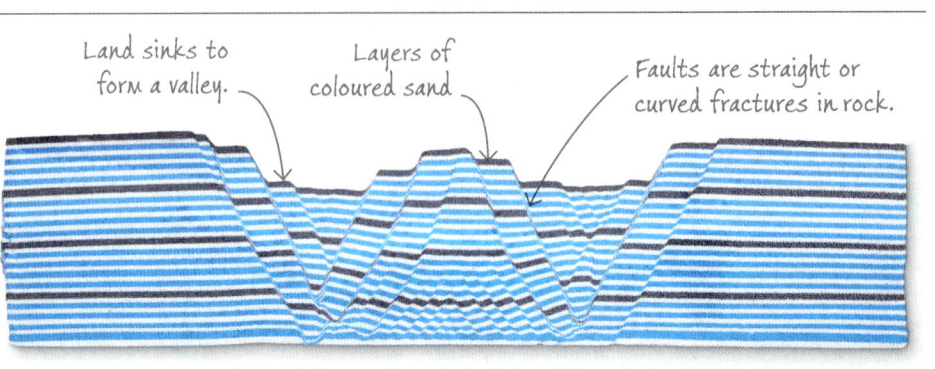

Land sinks to form a valley.

Layers of coloured sand

Faults are straight or curved fractures in rock.

The lowest city on Earth is Jericho near the Dead Sea.

The red pins are volcanoes that have been active in the last 10,000 years

The Great Rift Valley is splitting Africa apart.

Mount Kilimanjaro is a dormant stratovolcano.

DEAD SEA
Located between Israel and Jordan, the Dead Sea is a salt lake that is part of the Jordan Rift Valley. Its surface is 429 m (1,407 ft) below sea level, and its shoreline is the lowest land on Earth.

ERTA ALE
Numerous volcanoes are found along the Great Rift Valley, including Erta Ale in Ethiopia, which is a shield volcano with a lava lake that has been active for around 90 years.

The lowest point in Africa is Lake Assal in Djibouti.

LAKE NAKURU
Low points in the Great Rift Valley have flooded to form lakes. Lake Nakuru in Kenya is a soda lake, which means it is rich in alkaline salts. It's famous for the large flocks of flamingos that feed on the algae in the water.

HELL'S GATE
Hell's Gate in Kenya is known for its geothermal activity and for the spectacular Ol Njorowa Gorge, which was created by a river flowing out of Lake Naivasha, one of the Rift Valley lakes.

HOW WEATHERING WORKS

Some rocks are harder than others, but every kind of rock is eventually broken down by a process called weathering. Pattering rain, gusts of wind, and even your feet stomping along a mountain path all cause weathering. This process wears down rocks into tiny particles such as sand grains, while another process – erosion – carries these away. Working together over millions of years, weathering and erosion can wear away whole mountain ranges.

▼ WEATHERED GRANITE

This granite outcrop on a hill in Dartmoor, UK, is slowly wearing away due to weathering. Granite is a very hard rock made of the minerals feldspar, biotite, and quartz. Although tough, it eventually breaks down. Feldspar and biotite react chemically with the natural acidity in rain and turn into clay, which is softer, causing the rock to crumble. Quartz is much tougher but its crystals fall out as sand grains when the other minerals in granite are weathered away.

FERTILE SOIL

Soil is made from weathered particles of rock mixed with plant and animal remains. Soils made from a mix of rock types contain lots of different minerals. These soils are often the most fertile, making them perfect for growing crops.

BIOLOGICAL WEATHERING

Weathering happens in three different ways. When living things break up rock, it's called biological weathering. For example, pieces of rock may be broken off when a plant's roots grow into a crack in a rock and widen it until the rock splits.

PHYSICAL WEATHERING

When rocks are broken down by physical processes, it's called physical weathering. This rock in Antarctica has been split up by a process called freeze-thaw. Water trickles into cracks in a rock and expands when it freezes, causing the cracks to widen until the rock splits.

CHEMICAL WEATHERING

Chemical weathering occurs when rocks are broken down by chemical reactions. Rain picks up carbon dioxide from the air, making it slightly acidic. It trickles into cracks, where a chemical reaction occurs, changing and weakening some of the minerals.

Cracks in the rock allow rain to penetrate, causing chemical weathering.

SHRINKING MOUNTAINS

The inland mountain ranges of the western USA, such as the White Mountains, were once as high as the Himalayas, with peaks rivalling Mount Everest. After millions of years of weathering, the tallest peak in this region – White Mountain in California – is now 4,344 m (14,252 ft) – less than half of Everest's height.

BREAKING UP

Solid rock falls apart and breaks down into its mineral grains when weathered. Fragments of broken rock are carried away by rivers and dumped far away, even in the sea.

HOW EROSION WORKS

Earth's landscapes are constantly changing due to erosion – the wearing away of rock and the removal of rock particles by wind, water, and gravity. Working together with weathering (the breakdown of rock into smaller particles), erosion can create spectacular landforms, from cliffs and canyons to rock arches and pinnacles.

▶ ROCK PINNACLE
Pinnacles of rock, like this one in the Sahara desert, are shaped by erosion. This pinnacle has been worn away by windblown sand. Its tapering shape shows that the force of erosion is most powerful near the base, where a vortex of swirling wind has also made a depression in the sand. Eventually the pinnacle will become unstable and collapse.

SAND
Look closely at sand and you'll see it's made of tiny crystals. Most types of sand consist of grains of quartz – a hard, crystalline mineral found in igneous rocks like granite. When rocks break down, the quartz crystals fall out and are carried away by wind or water.

Quartz crystal

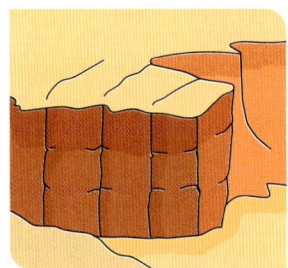

1 WALL
The side of a plateau (a raised area of land) is weathered and eroded by water and wind, leaving a narrow wall of rock.

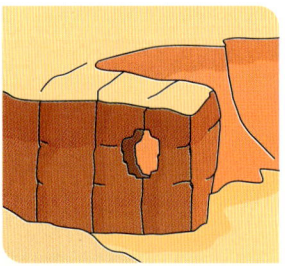

2 WINDOW
The wall becomes thinner as weathering and erosion continue. Windows appear in the softer, weaker areas of the rock wall.

3 PINNACLE
Eventually, a column of rock, known as a pinnacle, is all that remains of the wall. In time, the pinnacle will be eroded away too.

EROSION BY WIND
The sand-filled winds of the Sahara erode exposed soft rock into intricate patterns, such as this latticework of sandstone in Tassili n'Ajjer, Algeria.

EROSION BY RIVERS
Over millions of years, the Colorado River cut into the rocky ground to form this sharp turn, known as Horseshoe Bend, in Arizona, USA.

EROSION BY WAVES
The pounding of ocean waves can erode cliffs, creating rock formations such as this sea arch, known as Durdle Door, in the UK.

EROSION BY ICE
Ice erodes land too. Glaciers, such as Elephant Foot in Greenland, drag rocks and grit downhill, reshaping valleys and creating new landforms.

The sides of this pinnacle are slowly eaten away by windblown sand.

The depression is due to wind swirling around the base. The same thing happens when a hollow forms in the snow around a fencepost.

HOW LANDSLIDES WORK

When large amounts of earth, rock, and soil move down a steep slope under the force of gravity we call it a landslide. Every year, this type of erosion causes thousands of deaths and billions of pounds' worth of damage. Most landslides are triggered by rainfall, which soaks into the soil making it heavier, weaker, and more slippery. Shaking from earthquakes, volcanic rumbles, and human activities such as road cutting can also start rocks and soil tumbling.

Bare slope once covered in soil and forest

Houses engulfed by soil and uprooted trees

▶ AFTER AN EARTHQUAKE

In 2018, a powerful earthquake shook the island of Hokkaido in Japan, triggering landslides that killed 36 people. To make matters worse, heavy rains before the earthquake had loosened the soil covering the hills, making it more likely to slide downhill.

TYPES OF LANDSLIDE

Landslides can be fast or slow. They may be noticeable only after many years as rock and soil gradually move downhill. Others are much more dramatic as earth and rocks break away from a slope, forming a chaotic mixture of debris.

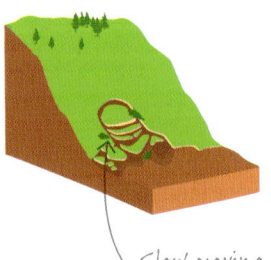
Slow-moving soil

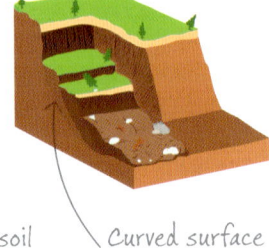
Curved surface

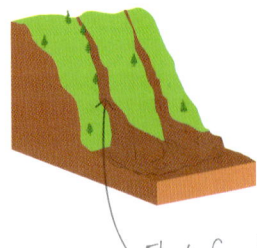
Flow of mud

Rock fragments

CREEP
This is the slowest type of landslide. Dislodged soil gradually slides downhill and may show as a wavy surface on a hillside.

SLUMP
In this type of landslide, part of a mountainside or cliff detaches from the bedrock and slides downwards, often along a curved surface.

MUDFLOW
Mudflows occur when heavy rain or melting snow turns soil into runny mud that surges down a slope in channels.

ROCKFALL
Rocks broken up by weathering may tumble down cliffs and mountainsides. A pile of fallen rocks is known as scree.

LAHARS

Volcanic eruptions can cause deadly mudflows, called lahars, that race downhill and can bury settlements. Lahars happen when an eruption melts snow or ice at the top of a volcano or when torrential rain mixes with the loose volcanic debris.

LAHAR SEDIMENT AT MOUNT ST HELENS, USA

RECEDING CLIFFS

Ocean waves pound away at cliffs with great power, weakening them and causing landslides. Over time, coastal cliffs recede, threatening communities that were once a safe distance from the sea.

COASTAL EROSION IN CALIFORNIA, USA

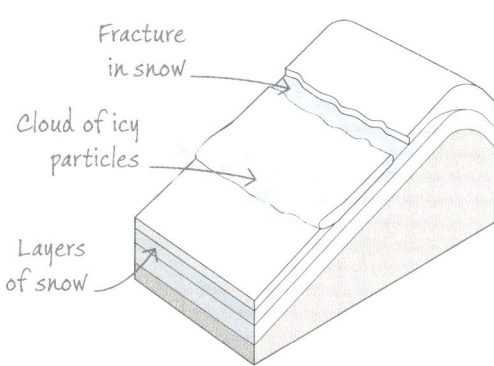

Fracture in snow

Cloud of icy particles

Layers of snow

AVALANCHES

Avalanches are "landslides" of snow. They occur when a recent layer of snow on a steep slope separates and slides away from the older layers underneath, gathering speed as it falls and creating a cloud of icy particles.

TYPES OF SAND DUNE

The shape and size of sand dunes depend on how fast and strong the wind is, which direction it comes from, and how much sand it carries. There are five different types of sand dune.

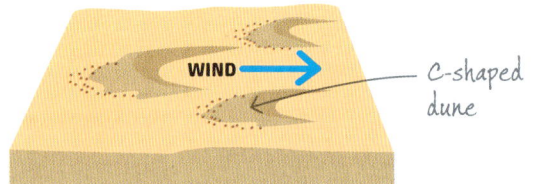

WIND →

C-shaped dune

Barchan dunes
These C-shaped dunes are the most common type of dune. They form in places where the wind usually blows from one direction.

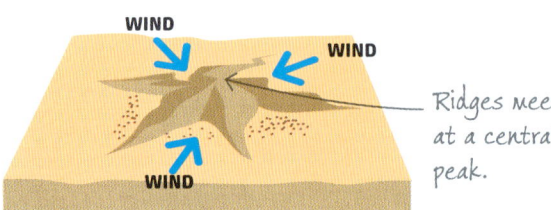

WIND
WIND
WIND

Ridges meet at a central peak.

Star dunes
Many of the tallest dunes are star dunes. They have three or more ridges and form where the wind blows from different directions.

HOW SAND DUNES WORK

Sand dunes are ever-changing mountains of sand, created and shaped by the wind. They are common on windswept beaches, but the largest occur in deserts, where vast expanses of sand dunes – sand seas – can stretch for hundreds of miles.

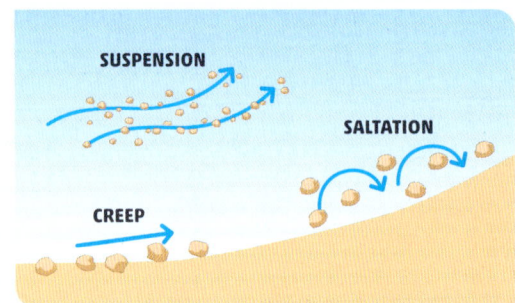

SUSPENSION

SALTATION

CREEP

HOW PARTICLES MOVE

Wind moves sand grains in three different ways. Large sand particles roll along the ground (creep). Medium particles move in hops and skips (saltation), and the finest particles become airborne (suspension). The faster the wind, the larger the particles it can lift and carry.

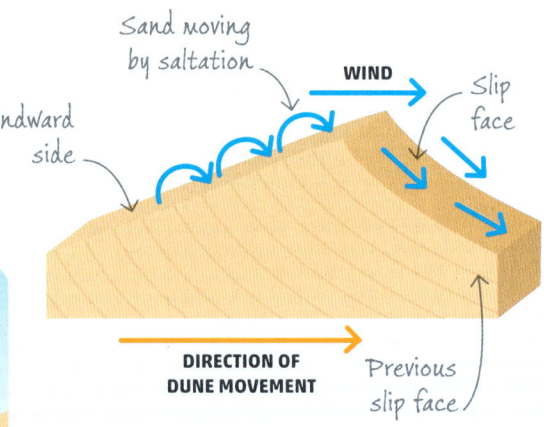

Sand moving by saltation

WIND →

Slip face

Windward side

DIRECTION OF DUNE MOVEMENT →

Previous slip face

MOVING DUNES

A dune begins to form when an obstacle slows the wind, causing it to lose energy and dump its cargo of sand. The growing dune intercepts the wind, trapping yet more sand and so getting bigger. On the windward side, sand grains are blown up the shallow slope. They pile up at the top until the crest becomes unstable and collapses down the steep side. Each time this happens, the dune moves a little. Over a year, a barchan sand dune can migrate up to 100 m (330 ft).

Sand is blown onto the windward face of a sand dune.

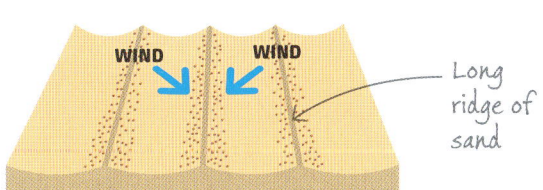

Linear dunes
These very long dunes sometimes stretch to 200 km (125 miles). They form where the wind varies between two directions.

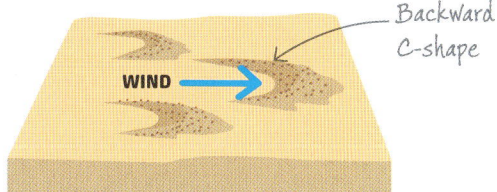

Parabolic dunes
Parabolic dunes form where the wind blows mainly from one direction. Their ends are held down by plants growing on the dune.

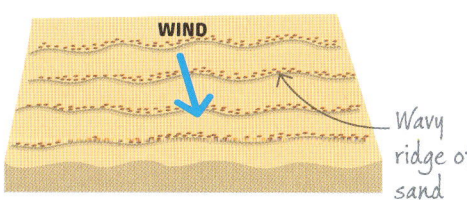

Transverse dunes
These are rows of very long, wavy dunes made when the wind blows from one direction and carries a lot of sand.

LIFE IN SAND DUNES
Sand dunes are difficult places to live, but some species have adapted to life here. The sandfish is a lizard that "swims" into the sand to avoid the Sun's heat. It tucks its limbs close to its smooth, streamlined body and slithers under the surface.

SINGING SAND DUNES
Sometimes a sand dune can make a low rumbling noise, like the hum of an aircraft. These sand-dune songs are caused by sand avalanches, which may be triggered by people, and can last for minutes and be heard miles away.

MARTIAN DUNES
Sand dunes form on other planets too. This false-coloured image shows crescent dunes (blue) near the north pole of Mars. The dunes are made of volcanic sand, and their peaks poke through a layer of carbon dioxide frost (pale areas).

▼ RUSTY RED DUNES
The wonderful red Soussusvlei sand dunes in Namibia, Africa, formed over many millions of years. They contain some of the tallest dunes in the world – similar in height to the Eiffel Tower in France. Their rusty colour comes from iron oxide in the sand, with the oldest dunes having the most intense red hue.

The slip face of a sand dune is the side that's sheltered from the wind.

HOW GLACIERS WORK

Glaciers are giant, flowing bodies of ice, like very slow-flowing rivers. They form in cold places where snow builds up faster than it melts, such as mountains and polar regions. Most glaciers move less than 1 m (3 ft) a day, but over time they drastically change the shape of landscapes, grinding down rock and gouging deep valleys between mountains.

GLACIER CAVES

Meltwater seeps down through crevasses and erodes ice until it reaches the base of the glacier, forming tunnels and caves. The Breidamerkurjökull caves in Iceland have a crystal blue colour because the glacial ice filters out all colours of light except blue.

CREVASSES

Different sections of a glacier move at different rates, resulting in the formation of giant cracks called crevasses. These deep openings, with near-vertical sides, are a hazard to skiers and mountaineers.

▼ ALETSCH GLACIER

Stretching for over 20 km (12.5 miles), and with a maximum depth of 900 m (2,950 ft), the Aletsch Glacier in Switzerland is the largest glacier in the European Alps. This valley glacier has been gouging out a U-shaped valley for thousands of years but, like many glaciers, it is now shrinking due to climate change.

VALLEY GLACIER
Valley glaciers are giant rivers of ice that get trapped between mountains as they flow downhill.

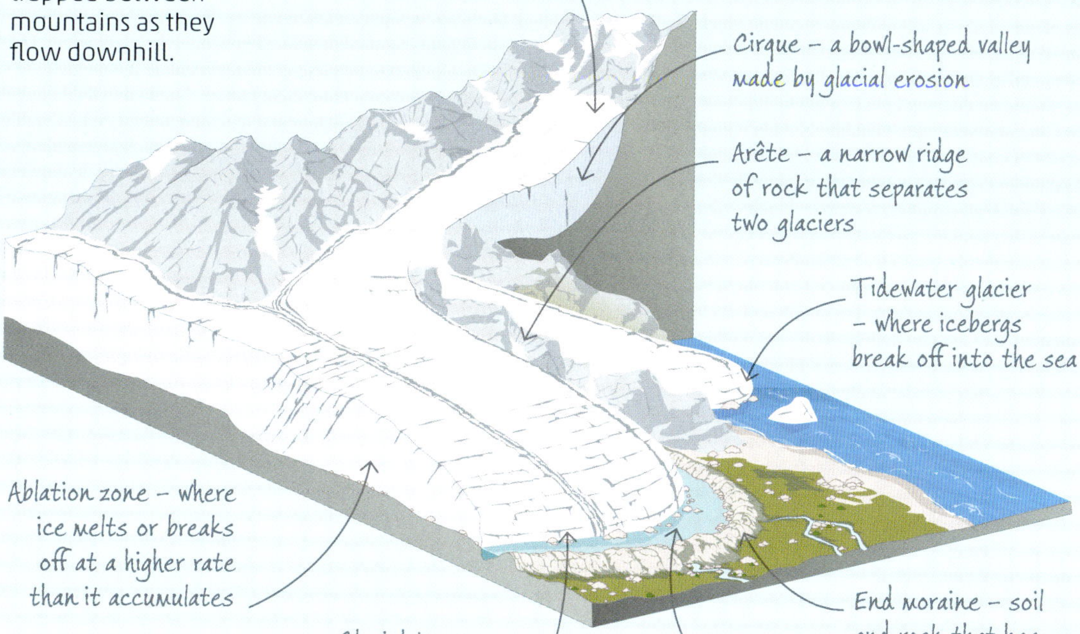

Accumulation zone – where snowfall is greater than ice loss

Cirque – a bowl-shaped valley made by glacial erosion

Arête – a narrow ridge of rock that separates two glaciers

Tidewater glacier – where icebergs break off into the sea

Ablation zone – where ice melts or breaks off at a higher rate than it accumulates

Glacial toe or terminus – where the glacier ends

Meltwater stream

End moraine – soil and rock that has been picked up by the glacier and dumped at its end

When two glaciers meet, soil and rock debris from their edges combine to form a central streak – a medial moraine.

FORMATION OF GLACIAL ICE
Glaciers form when snow builds up in the same place. Over time, the snow is squashed by the weight of new snow, squeezing out air. This turns it into a denser, granular substance called firn, which resembles wet sugar, and finally into solid ice. It can take thousands of years for ice to reach the bottom of a glacier, and scientists estimate that ice at the bottom of the Antarctic ice sheet could be 1 million years old.

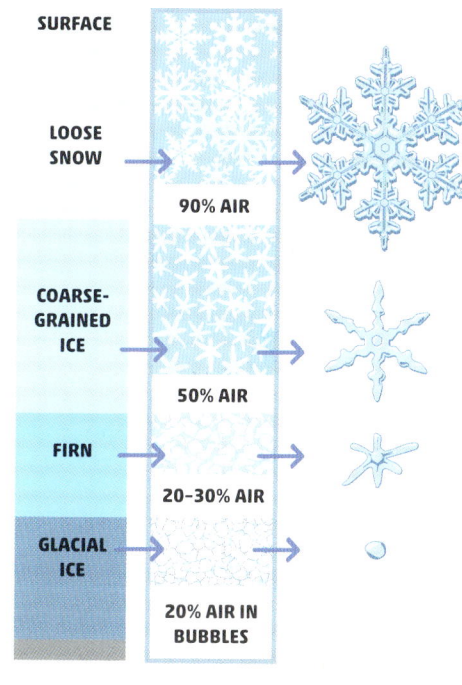

SURFACE

LOOSE SNOW

90% AIR

COARSE-GRAINED ICE

50% AIR

FIRN

20–30% AIR

GLACIAL ICE

20% AIR IN BUBBLES

GLACIAL VALLEYS
When glaciers melt, they leave behind U-shaped valleys with steep sides and a curved floor. These are different from the V-shaped valleys carved out by rivers, because erosion by the glacier happens across the whole valley, widening it and making the sides steeper. There are many glacial valleys in places that used to be covered in glaciers during the Ice Age – such as Yosemite National Park, USA, and the fjords of Norway.

GLACIER   **U-SHAPED VALLEY**

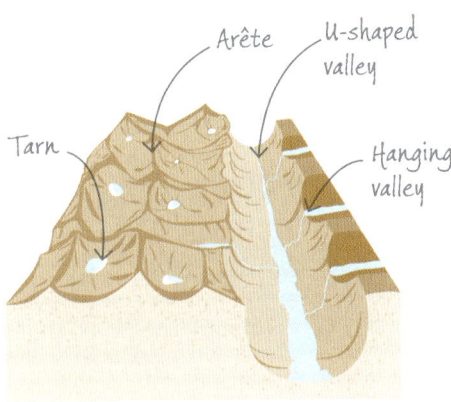

CARVED-OUT LANDSCAPE

In places once covered in glaciers, telltale signs remain showing that the rivers of ice carved mountains and valleys into different shapes.

HANGING VALLEY

A hanging valley is a high side valley created by a tributary glacier that once fed into a larger glacier. The valley is higher because the smaller glacier contained less ice and didn't erode as deeply as the main glacier. Hanging valleys often end in waterfalls.

◀ STIRLING FALLS, NEW ZEALAND

ARÊTE

An arête is a sharp, rocky ridge that was carved between neighbouring glacial valleys. The name comes from the French word for a fish's backbone, which the ridge resembles.

◀ BEN NEVIS, SCOTLAND

▲ GRANITE CREEK TARN, ALASKA

U-SHAPED VALLEY

As glaciers flow, they pick up rocks, which then grind away at V-shaped river valleys. This process results in U-shaped valleys, with steep sides and a rounded bottom.

YOSEMITE NATIONAL PARK, USA ▶

TARN

Small lakes that form high in the mountains, in the bowl-shaped valley (cirque) once filled by the head of a glacier, are called tarns. The word comes from the old Norse *tjörn*, which means pond.

GLACIAL FEATURES

Glaciers shape and mould our landscapes by carving out valleys, reshaping mountains, and shifting rocks and soil across huge distances. Glaciers once covered more than one-third of Earth's surface, and the features they left behind give us valuable information about past climates.

ERRATIC

Erratics are rocks carried by glaciers and dumped in a location where the local rock is different. They range from the size of a pebble to larger than a house. By studying erratics, geologists can work out where ancient glaciers once flowed.

YORKSHIRE, UK ▶

MORAINE

This is the name given to rocks and soil that are moved and then dumped by a glacier. The sediment that heaps up at the furthest reach of a glacier is called a terminal moraine and is a bit like the high-tide line on a beach.

Terminal moraine

◄ KASKAPAKTE GLACIER, SWEDEN

KETTLE LAKE

When big blocks of ice from retreating glaciers become stranded in the rocks and stones transported by the glaciers, they melt and create hollows in the ground. These hollows may later fill with water to form deep ponds called kettle lakes.

ALASKA, USA ▶

ESKER

Eskers are winding ridges of sand and gravel deposited by meltwater rivers that once flowed through tunnels under glaciers. The longest eskers stretch for hundreds of miles, and the tallest are 30 m (98 ft) high.

▼ MANITOBA, CANADA

DRUMLIN

As glaciers slide across the landscape, they create new lumps and bumps, known as drumlins, from the debris they carry. The mounds are tapered on one side, pointing in the direction that the glacier once flowed in.

◄ CLEW BAY, IRELAND

GLACIAL DEPOSITS

Glaciers transport rocks and soil as they flow. When glaciers melt, this debris is left behind, forming distinctive landscape features such as drumlins, eskers, erratics, and moraines.

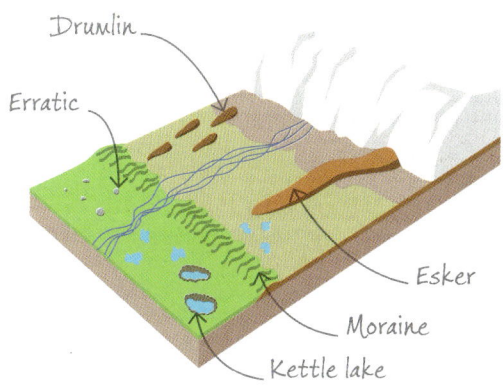

Drumlin

Erratic

Esker

Moraine

Kettle lake

HOW
ICEBERGS
WORK

Icebergs are like giant ice cubes floating in the ocean. They are made of freshwater ice that has broken off the end of a glacier or an ice shelf. Formed in the coldest parts of the world, icebergs are found in the Arctic and North Atlantic oceans and in the oceans around Antarctica. Icebergs may drift thousands of miles on ocean currents and take several years to melt.

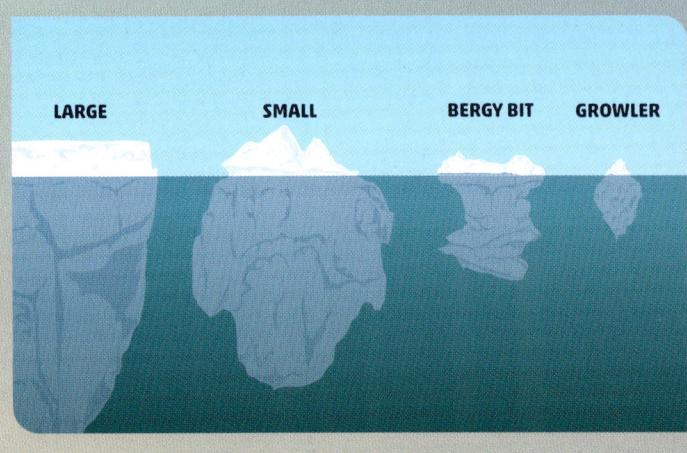

ICEBERGS, BERGY BITS, AND GROWLERS
Icebergs are classified by their size. The smallest ones, growlers, are about the size of a car. Bergy bits, next up in size, are roughly as big as a small house. The other four categories are simply called small, medium, large, and very large. The biggest icebergs of all are larger than some countries.

UNDERWATER VIEW
Seen from under water, the iceberg has a mottled appearance because melting ice leaves pockmarks in its surface. Rock fragments and dirt picked up by the glacier's base on land sink to the sea floor as the iceberg melts, providing nutrients for marine life.

Tiny air bubbles inside ice reflect white light, giving icebergs their white appearance.

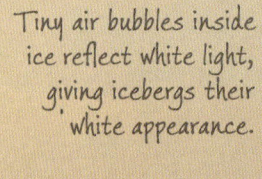

GIANT ICEBERGS
Rifts (cracks) in ice shelves around Antarctica create the most gigantic icebergs, which are flat in shape (tabular). In 2017, this rift in the Larsen C Ice Shelf allowed a vast iceberg, known as A-68, to break away. It was 175 km (110 miles) long and 50 km (30 miles) wide.

WHERE DO ICEBERGS COME FROM?
When icebergs form – a process called calving – they create a loud cracking or booming noise as the ice hits the sea. They can also cause large waves. Most icebergs in the northern hemisphere come from Greenland, with the West Greenland glaciers calving 10,000 or more icebergs every year. Most icebergs in the southern hemisphere come from Antarctica.

ICE SCULPTURES
Icebergs don't melt evenly all over. The ceaseless action of wind, rain, waves, and ocean currents can erode icebergs unevenly, sometimes carving them into fantastic shapes, such as this arched iceberg in the North Atlantic Ocean.

The blue parts of icebergs are made of ice with fewer air bubbles.

◄ FLOATING ISLANDS
Ice is less dense than water, which is why ice cubes float in your drink and icebergs float in the sea. Very large icebergs can weigh more than 10 million tonnes and tower as high as 20 double-decker buses over the surface of the ocean. Most of an iceberg is hidden under water, with only about one-tenth of its volume poking up above the waves.

HOW **RIVERS** WORK

Rain and melting snow flow off the land and drain away in rivers. As well as transporting water, rivers sweep away all sorts of sediment – rocks, pebbles, gravel, sand, silt, and clay. Over great spans of time, this slow but relentless process transforms landscapes. It carves out V-shaped valleys in the upper and middle parts of rivers, and it creates flat plains where sediment is dumped in the lowlands further downstream.

▼ DRAINAGE BASIN
A river doesn't have a single source. It is fed by many streams and by groundwater from a large area called a drainage basin. Rivers flow downhill all the way, often starting in mountains. The upper parts are steep, with fast-flowing water and rocky rapids. The middle parts pass through wider valleys formed by erosion over tens of thousands of years. Finally, the river, now flatter and calmer, snakes in wide curves across broad plains.

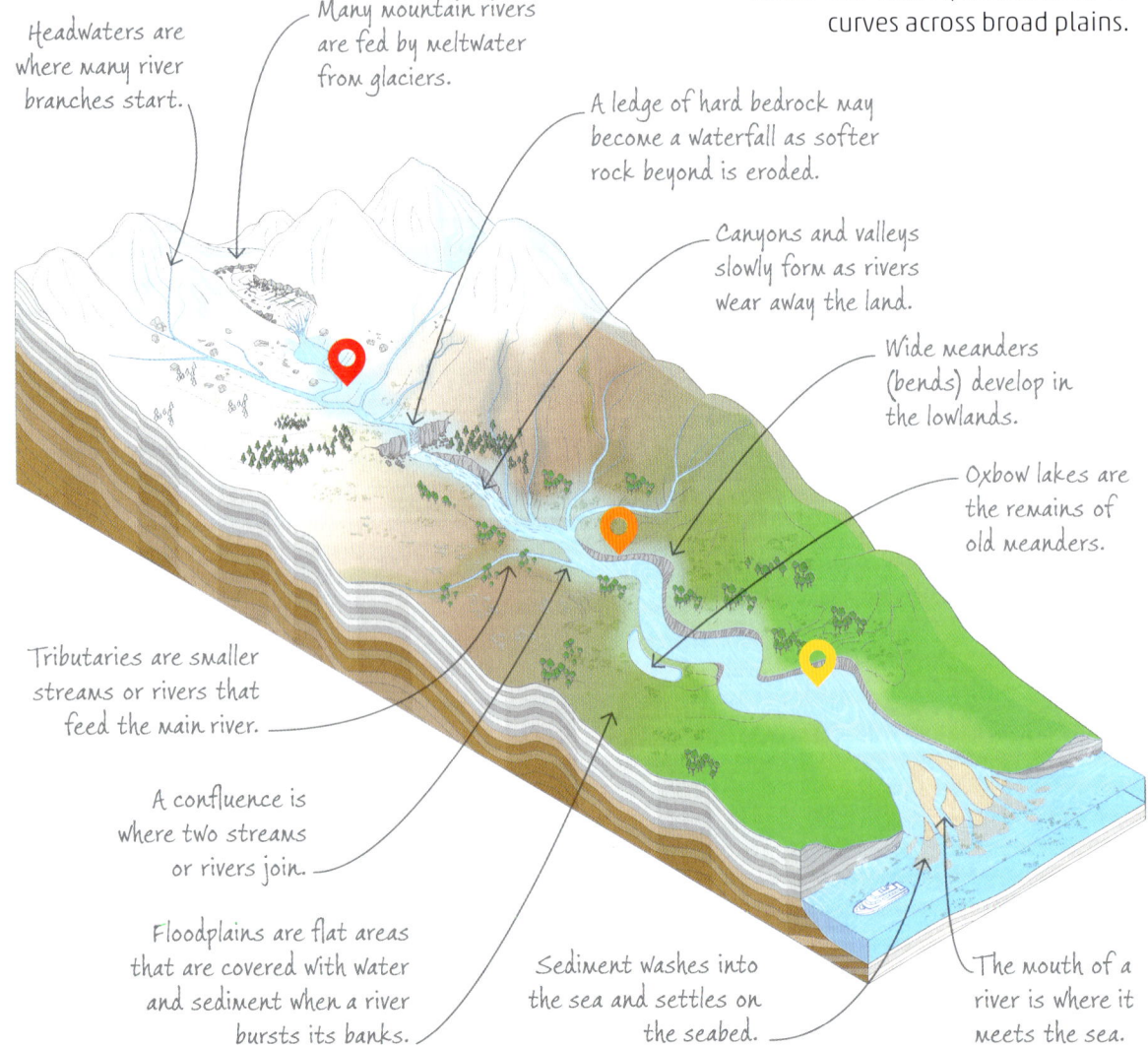

Headwaters are where many river branches start.

Many mountain rivers are fed by meltwater from glaciers.

A ledge of hard bedrock may become a waterfall as softer rock beyond is eroded.

Canyons and valleys slowly form as rivers wear away the land.

Wide meanders (bends) develop in the lowlands.

Oxbow lakes are the remains of old meanders.

Tributaries are smaller streams or rivers that feed the main river.

A confluence is where two streams or rivers join.

Floodplains are flat areas that are covered with water and sediment when a river bursts its banks.

Sediment washes into the sea and settles on the seabed.

The mouth of a river is where it meets the sea.

HIGHLANDS

In mountains, fast-flowing streams called rapids cut vertically into the landscape, carving out V-shaped valleys. The uneven, steep ground makes the water turbulent. The steep hillsides cause rockfalls that send boulders tumbling into the stream. As the water flows around these obstacles, it forms powerful currents.

TRANSITION ZONE

Swollen by water from many tributaries, rivers grow larger as they leave mountain valleys. Broader valleys form in the middle sections of a river under the force of the water flowing over thousands of years, and millions of rocks and pebbles are dumped on the riverbed as the water becomes less turbulent.

LOWLANDS

The lower part of a river meanders in wide loops over flatter land. These bends continually change as sediment is eroded in some places but dumped in others. Occasional floods spread sediment over large floodplains. As the river slowly cuts down into the land, former floodplains are stranded as flat areas of land (terraces).

HOW RIVERS BEND

All natural rivers snake across the land with a wavy, bendy pattern. The bends in rivers are not fixed. They continually change as rivers erode the ground, shifting vast amounts of soil, sand, and pebbles from place to place. Over long periods, rivers change course, but dramatic changes can also happen suddenly during floods.

Water supply
Sediment
Sloping table
Filter removes sediment
Reservoir and pump
Water outlet

RIVER TABLE
Scientists study how rivers change over time by modelling them in a river table. Water is pumped to the upper end of the table and allowed to flow through a bed of artificial sediment. The flowing water washes sediment downstream and carves a channel that continually changes shape.

▼ MODELLING RIVERS
It can take many years for large rivers to change course, but the same process takes minutes when a river is modelled in a lab. The water picks up sediment where it flows quickly and deposits it where it slows down. Bends grow larger over time because water flows fastest round the outside of a bend, eroding the ground there more powerfully.

Sediment eroded

Sediment deposited

Bend develops

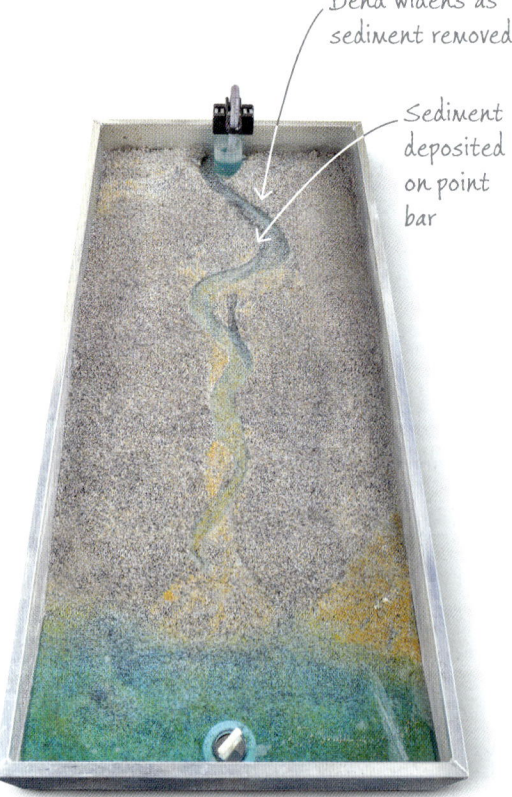

Bend widens as sediment removed

Sediment deposited on point bar

1 CHANNEL FORMS
Flowing water creates a channel as it removes sediment. The most lightweight particles (yellow) are picked up first and deposited downstream in slower water.

2 BENDS FORM
Bends can form wherever small obstacles divert the water to one side. Water flows faster around the outsides of bends, so the bends get progressively larger over time.

3 BENDS WIDEN
A succession of bends develops, giving the river a wavy shape. Sediment builds up on the insides of bends, where water is slower, forming shallow banks called point bars. The river flows around these obstacles, making the bends even wider.

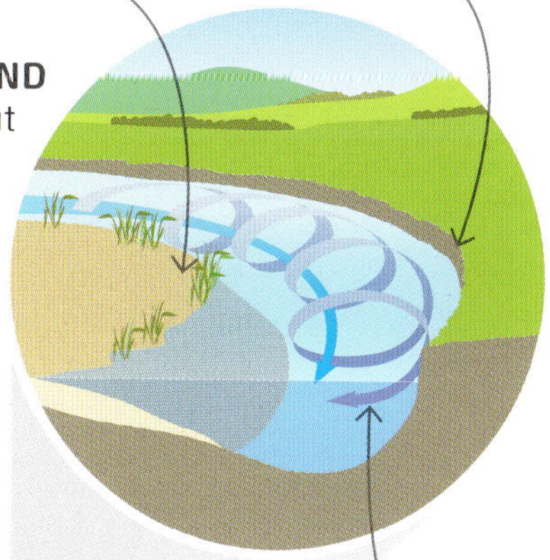

Gentle river bank where sediment is deposited on point bar

Steep eroding river bank

INSIDE A BEND
The river is not only faster but deeper in the outside of a bend. The water spirals around as it rushes downstream, cutting out a steeply eroded river bank. On the opposite, slower side of the river, sediment settles out to form a shallow bank of sand and pebbles – a point bar.

Flow spirals around

Shortcut forming

4 **SHORTCUTS FORM**
A bend may get so large that the river finds a shortcut. This can lead to the formation of an oxbow lake – a stretch of water cut off from the river.

OXBOW LAKES
River bends get larger and larger over time. Eventually only a narrow neck of land is left between neighbouring loops. If the river breaks through this, the new channel has faster flow and cuts out a new riverbed, leaving sediment at the side. This cuts off the old loop, creating an oxbow lake.

BENDS WIDEN   **LOOP FORMS**   **OXBOW LAKE**

POOLS AND RIFFLES
Even a single pebble can give birth to a river bend. The tiny disturbance to water flow around a pebble causes changes that are magnified over time, creating features such as pools, riffles (submerged pebble banks), and bends.

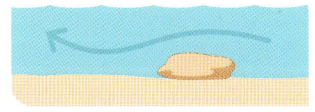

1 **OBSTRUCTION**
When water meets a pebble on the riverbed, the flow is squeezed over the top.

2 **HOLLOWS FORM**
As water swirls around the pebble, it speeds up and collects sand grains from the riverbed. Hollows form where sand is removed.

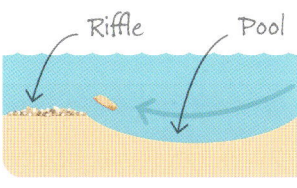

Riffle   Pool

3 **POOL AND RIFFLE FORM**
The pebble is swept away and the hollows merge and grow to form a pool. Downstream is a riffle, where pebbles and sand are dumped.

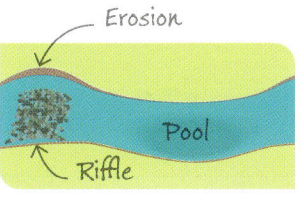

Erosion

Pool

Riffle

4 **BEND FORMS**
The riffle is an obstacle, so water tries to flow around it, causing the river bank to erode and a bend to form.

BRAIDED RIVER

Most rivers form a single channel. However, if the water volume is large, the slope steep, or the water is choked with fine sediment, a river may break up into a network of interweaving channels – a braided river. Like this one in Iceland, braided rivers have dozens of muddy islands where sediment collects. The islands are unstable. They expand, shrink, and vanish as the channels continually change course.

HOW
WATERFALLS
WORK

A waterfall shows the power of erosion in action. The plunging water and its cargo of pebbles continually pound the riverbed under the falls, gouging away bedrock and eating into the cliff. Waterfalls may look permanent, but they slowly inch their way upstream. Over thousands of years, all waterfalls eventually wear themselves away.

A sudden drop in a riverbed is called a knickpoint.

A hard rock overhang forms at the top of the waterfall.

Weaker rock is worn down by the force of water.

SEGMENTED WATERFALL
There are many different kinds of waterfall. Iguaçu Falls in South America is a segmented waterfall, which means it consists of many channels separated by islands of rock.

BLOCK WATERFALL
Block waterfalls form on wide rivers. The water descends in a wide, uninterrupted curtain. Horseshoe Falls – one of the three waterfalls that form Niagara Falls on the US–Canada border – is a block waterfall.

CASCADE WATERFALLS
At a cascade waterfall, the river tumbles over a series of rocky steps. Detian Falls on the Vietnam–China border is a cascade waterfall.

FAN WATERFALLS
Fan waterfalls spread out horizontally on the rock as the water spills down. Union Falls in Montana, USA, is a fan waterfall.

◀ PLUNGE WATERFALL
Helmcken Falls in British Columbia, Canada, is a plunge waterfall. The water plunges over an overhanging ledge and falls vertically through the air, without touching the bedrock, until it hits the pool below.

HOW WATERFALLS FORM
Waterfalls are most common on steep ground, where rivers flow with greater force. Many begin to form where a band of hard rock meets a band of soft rock.

Hard rock

Soft rock

❶ BEDROCK CHANGE
A waterfall can begin to form if the river flows across a boundary between hard, erosion-resistant rock and a softer type of rock.

❷ DROP FORMS
The softer rock erodes faster than the hard rock, creating a drop. The falling water gains force, speeding up the process of erosion.

Overhang

❸ PLUNGE POOL
Plunging water and pebbles swirl around the base of the cliff, carving out a wide plunge pool. The cliff beneath the resistant rock recedes, creating an overhang.

❹ MOVING UPSTREAM
Eventually, the overhang collapses as it has nothing supporting it from underneath. The cliff continues to recede, and the process repeats, causing the waterfall to move upstream.

VICTORIA FALLS

In the language of Africa's Lozi people, Victoria Falls is the "smoke that thunders". Visitors can feel the thunderous noise in their feet before setting eyes on the waterfall – the largest in the world, twice as tall and wide as Niagara. It formed where a crack in the basalt bedrock created a weak area. Erosion turned the crack into a gorge that now swallows the Zambezi River – along with the occasional unfortunate crocodile or hippo.

HOW FLOODS WORK

Floods happen when large amounts of water submerge land that is normally dry. Some floods happen in minutes and take people by surprise, but others build up gradually over months. Floods can do devastating damage, but some bring benefits. River floods, for example, spread sediment over land and fertilize soil.

▶ RIVER FLOODS

Melting snow or heavy rain can make rivers overflow. In 2019, a particularly wet winter and spring in the midwestern USA led to catastrophic floods of the Mississippi River. Urban areas were submerged, causing loss of lives and billions of dollars' worth of damage. These satellite images show flooding near the city of Memphis, which is on the right.

FEBRUARY 2014 (BEFORE FLOOD)

FEBRUARY 2019 (DURING FLOOD)

FLOOD DEFENCES

While some floods are inevitable, there are ways to reduce the damage they cause or even prevent them altogether.

NATURAL DEFENCES

Many wetlands are great at absorbing excess water, and then releasing it slowly during dry seasons. Trees also absorb a lot of water through their roots, and densely vegetated ground slows the flow of water. Restoring and promoting natural habitats, for example by reintroducing beavers to an area, can therefore protect us against flooding.

ARTIFICIAL DEFENCES

Most artificial flood defences are physical barriers that hold back water, either inside a river or along its banks. The Thames Barrier in London, UK, protects the city from dangerously high tides by blocking the incoming water with rotating gates between concrete piers.

PLUVIAL FLOODS

When heavy rain hits hard ground in urban areas, drainage systems may be overwhelmed, causing pluvial (surface water) floods. Trapped by buildings and paved areas, the water pools in low spots, turning streets into temporary rivers.

COASTAL FLOODS

When high tides coincide with storms, large waves may flood coastal areas. Hurricanes can also raise the sea level, causing even bigger floods, and submarine earthquakes can cause tsunamis that sweep inland for miles.

GROUNDWATER FLOODS

Soil and porous rock layers underground absorb rainwater like a sponge. This groundwater normally drains away slowly towards rivers and the coast, but prolonged rain can raise it faster than it drains, until it appears above ground and causes flooding.

HOW **CANYONS** WORK

Canyons are deep, steep-sided valleys with rocky walls. They form over millions of years, as rivers slowly erode the rock and sink deeper into the ground. Most large canyons occur in hot, dry regions, where weathering and erosion mainly affect areas near the river. In wetter climates, rivers tend to create wider V-shaped valleys instead, as the weathering is more widespread.

▼ **GRAND CANYON**
The Grand Canyon in Arizona, USA, is so vast that if all of the water in all rivers on Earth was put into it, the canyon would only be half full. There are nearly 40 distinctive bands of rock that have been revealed on the sides of the canyon, with the oldest at the bottom dating back almost two billion years.

The surrounding landscape is flat and dry.

Bands of different rock types wear down at different rates, creating steps.

Fallen rocks build up in piles of rubble (scree) at the base of the cliffs.

The Colorado River slowly wears down rock on the bed of the river.

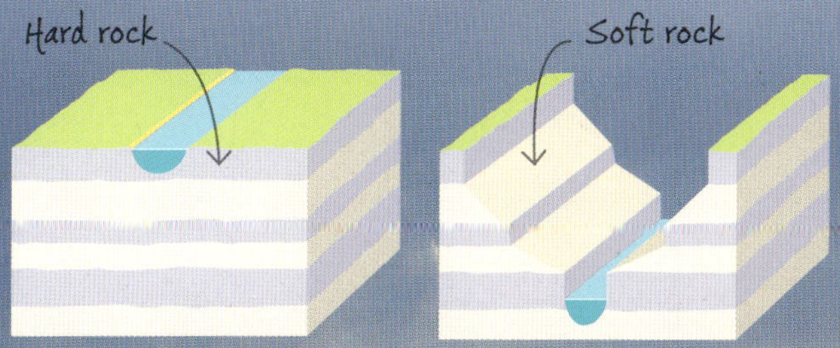

Hard rock   Soft rock

STAIR-STEP CANYON

HOW THE GRAND CANYON FORMED

The Grand Canyon is a stair-step canyon. Some parts began to form 70 million years ago, but most parts formed over the last 6 million years as the Colorado River carved down into the rock. The steps are formed because the Colorado plateau has alternating layers of soft and hard rock, which erode at different rates.

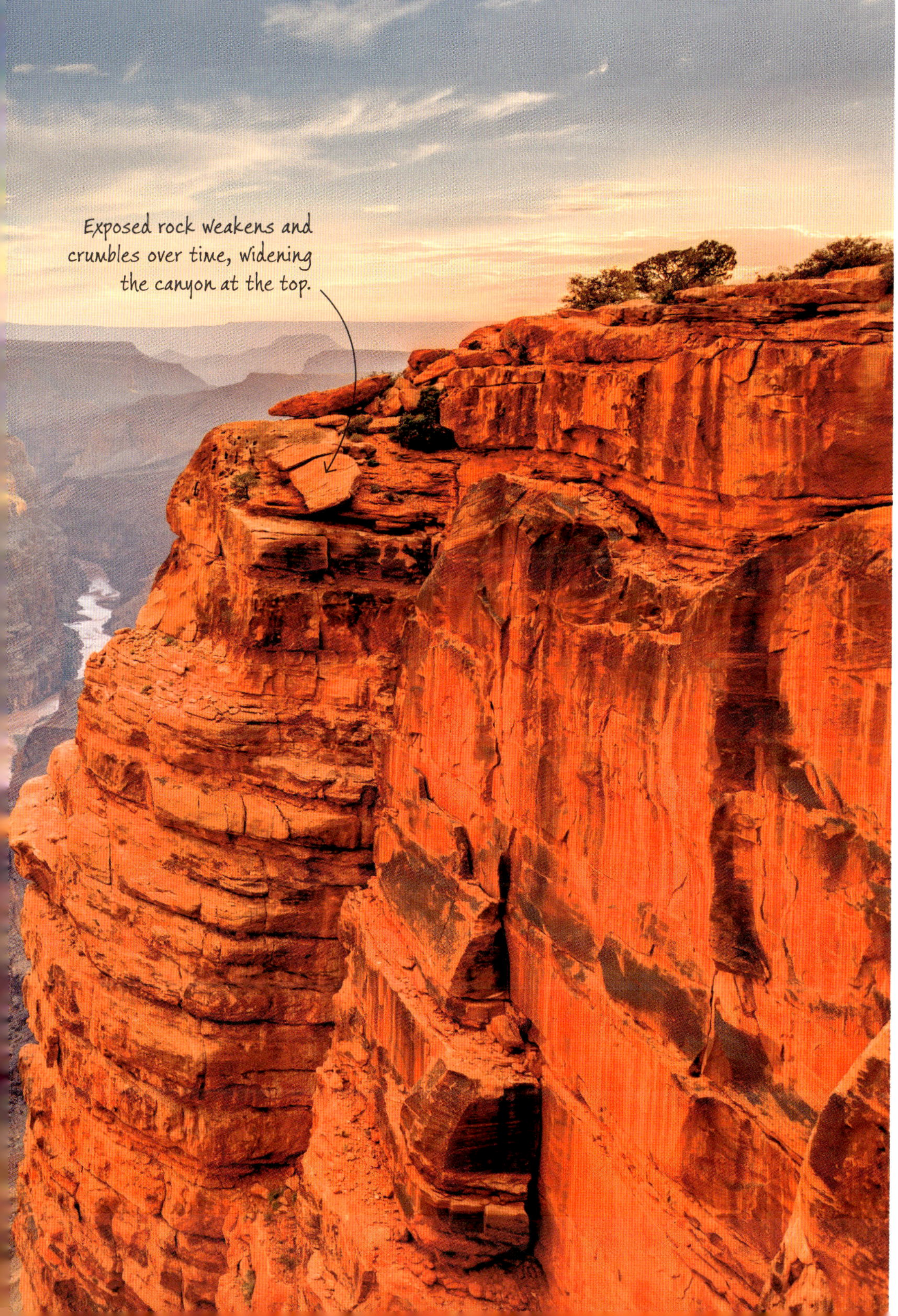

Exposed rock weakens and crumbles over time, widening the canyon at the top.

COLOURED CANYON, EGYPT

SLOT CANYONS

Narrow canyons with vertical walls, such as the Coloured Canyon in Egypt, form when water quickly cuts down through a single rock layer. The water erodes the rock through abrasion – it picks up debris and then acts like sandpaper against the canyon walls, particularly during flash floods.

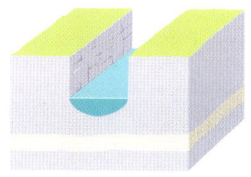

VERTICAL EROSION
Water carves quickly downwards, through a single rock layer.

FLOOR FORMATION
If water hits a softer rock layer, the canyon will suddenly widen.

COMPUTER-GENERATED DEPTH MAP OF MONTEREY CANYON, CALIFORNIA

CANYONS IN THE SEA

Submarine canyons are found on the sea bed near continents. Rivers may have cut the upper parts of these canyons long ago, when sea levels were much lower. When sea levels later rose, underwater landslides and ocean currents deepened and lengthened the canyons.

ANTELOPE CANYON

The flowing shapes in the walls of this canyon in Arizona, USA, are a clue to how it formed. Antelope Canyon is a slot canyon that was carved from the desert sandstone by floods. For thousands of years, repeated flash floods laden with sand and grit have whipped through the canyon in a raging torrent, scouring the walls. Even today, floods fed by rainfall miles away can fill the canyon without warning.

HOW GROUNDWATER WORKS

More than 99 per cent of Earth's unfrozen fresh water is hidden underground, trapped in the tiny gaps between rock and soil particles. Called groundwater, this water moves far more slowly than surface water and can stay underground for thousands or millions of years. It helps to keep rivers flowing in dry periods, and it provides more than a fifth of the world's population with water for farming and drinking.

▼ OASIS

In the Sahara Desert, low points in the land allow groundwater to emerge, forming oases. Lake Gaberoun in Libya is fed by water that has been trapped in rock below the dunes for thousands of years. Evaporation has made the lake's water too salty to drink, but desert travellers can reach drinkable water from nearby wells.

The water in Lake Gaberoun is five times saltier than sea water. Bathers can float in the salty water without swimming.

A sand sea (erg) surrounds the oasis.

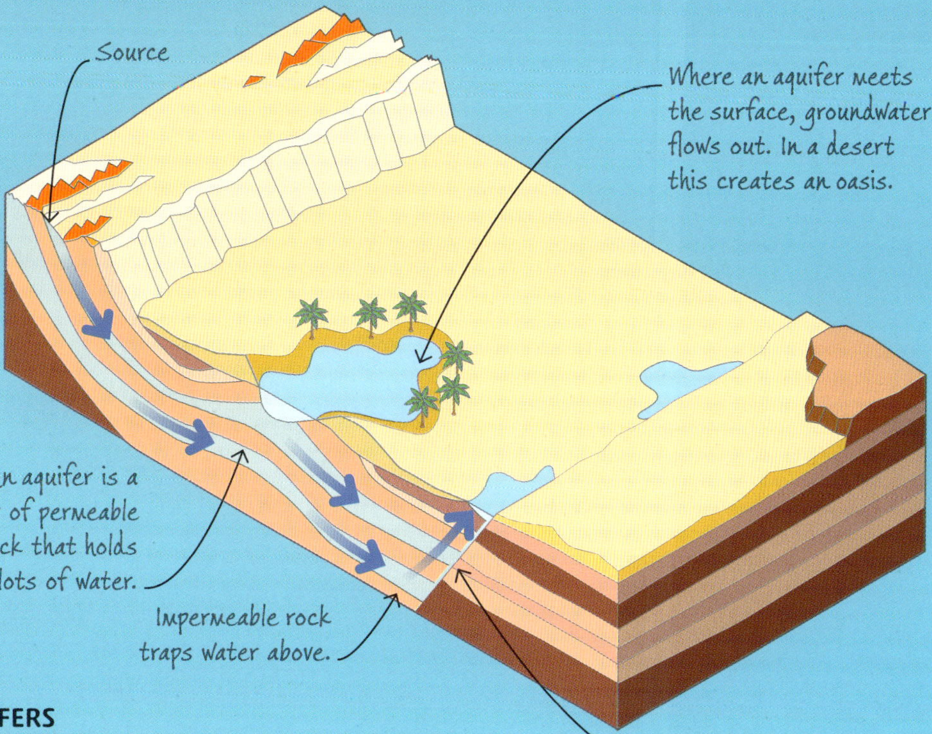

Source

Where an aquifer meets the surface, groundwater flows out. In a desert this creates an oasis.

An aquifer is a layer of permeable rock that holds lots of water.

Impermeable rock traps water above.

Faults (cracks) can force groundwater to the surface.

AQUIFERS

Some kinds of rock are permeable, which means that water can seep through them. If a layer of permeable rock lies over an impermeable layer, water builds up to form an aquifer – an underground water store. Vast amounts of water seep slowly through aquifers. If the aquifer meets the ground, the water flows out – often miles from the original source.

Date palms flourish around oases in the Sahara.

WATER IN ROCK

Certain kinds of rock have tiny gaps between the solid grains that water can seep into. For a rock to make a good aquifer, it must also have connections between the gaps so that water can flow. Sedimentary rocks such as sandstone and chalk make the best aquifers.

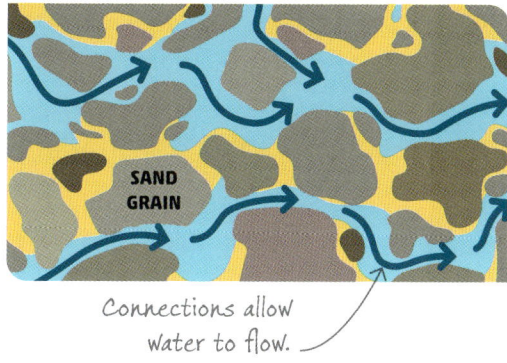

SAND GRAIN

Connections allow water to flow.

WELLS

Wells are deep holes that reach down to groundwater. Water usually has to be pumped out of a well or lifted with buckets. However, some wells have water that spurts upwards under its own pressure. These are called artesian wells and form when the hole taps into a confined aquifer that is fed by a water source higher than the well.

UNDERGROUND RIVERS

Not all groundwater is trapped within permeable rock. In limestone areas, the natural acidity in rain eats away at rock minerals to form cavities that eventually grow into caves. Underground rivers flow through these cave systems, in some places forming hidden waterfalls and lakes.

HOW CAVES WORK

Caves are natural underground spaces in rock that are large enough for people to explore. They vary in size from small hollows a person can barely squeeze into to vast caverns with connecting passages that are miles long. Some caves are full of water, while others are dry or dripping with moisture. Caves occur in several rock types, but most form in limestone.

Soda straws are delicate, hollow tubes formed by drips from the cave roof, each leaving a fine ring of the mineral calcite.

Stalactites form when soda straws get clogged and water starts running down the outside, allowing calcite to deposit in a thicker cone shape.

Columns are created when a stalactite and a stalagmite join together.

Stalagmites form from drips that hit the cave floor, building calcite deposits upwards over time.

Flowstones are sheets of calcite that form from water flowing down the walls or along the floor of a cave.

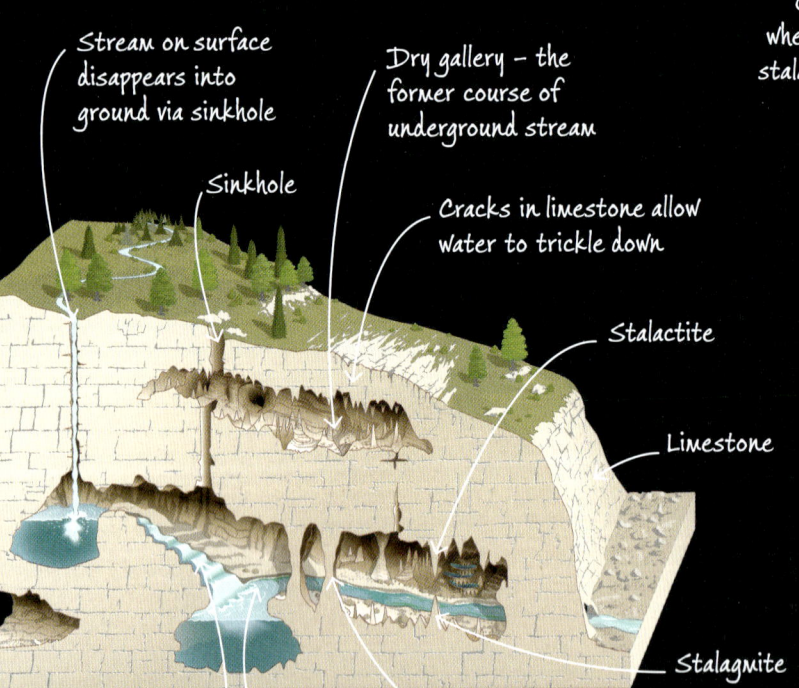

Stream on surface disappears into ground via sinkhole

Sinkhole

Dry gallery – the former course of underground stream

Cracks in limestone allow water to trickle down

Stalactite

Limestone

Stalagmite

Underground stream further enlarges cave system

Underground river

Column

LIMESTONE CAVE SYSTEM

Rain is naturally acidic because it absorbs carbon dioxide from the air. As a result, it reacts chemically with the minerals in limestone, dissolving them in some places to form caves and depositing them in others to form speleothems (stalactites, stalagmites, and so on).

◄ INSIDE A CAVE

Harrison's Cave is a limestone cave located in the Caribbean island of Barbados. It contains a variety of geological features, including stalactites, stalagmites, and columns. All these formations grow from mineral deposits left by the groundwater that trickles through the cave.

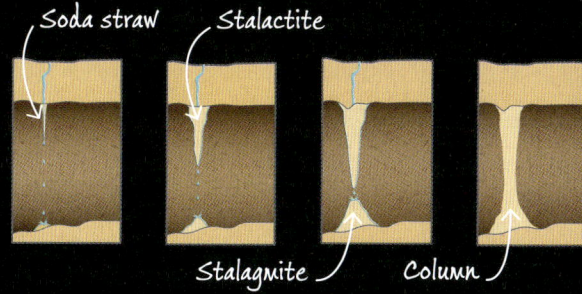

Soda straw   Stalactite

Stalagmite   Column

COLUMN FORMATION

Columns form when a stalactite and a stalagmite fuse. This process can take thousands of years. The tallest known column is just over 60 m (200 ft) and is located in a cave in Thailand.

FLOODED CAVES

Water can fill caves quickly when it rains. Cave explorers must check the weather forecast to avoid getting trapped, and many carry diving equipment so they can explore underwater.

CAVE ANIMALS

Many caves are home to their own unique wildlife. Eyes are not needed in the pitch darkness, which is why some species, such as this salamander, have lost them.

FERTILE GROUND

The ground in deltas is mostly silt – a type of sediment that makes rich, fertile soil. The Ganges Delta is one of the most fertile agricultural regions on Earth.

Rivers divide into many branches (distributaries) in a delta.

The dark green areas are natural mangrove forests.

HOW DELTAS FORM

Scientists study how deltas grow by running water through a large tray of artificial, coloured sand. The lightweight yellow sand is washed out to sea and builds up in a fan shape, forming new land. As sand piles up in one place, the river is forced to change course and so starts dumping sediment elsewhere. The delta expands, with the river continually changing course and branching as it flows though it.

1 EROSION
As a river flows, it erodes the ground and washes particles of sediment downstream towards the sea.

2 DEPOSITION
The river slows down as it nears the sea. This causes sediment particles to drop out of the water and build up.

3 FAN GROWS
The sediment forms a fan shape that extends into the sea, creating new land. The river changes position as it forces its way through.

4 DELTA
The river keeps moving and branching, forming new channels. Each channel deposits more sediment, enlarging the delta.

UNDER A DELTA

A delta grows because the river adds sediment faster than waves can erode it. The older sediments tend to be at the bottom and newer sediments on top. Pebbles and gravel build up where the river meets the sea, while finer sands and mud settle further out.

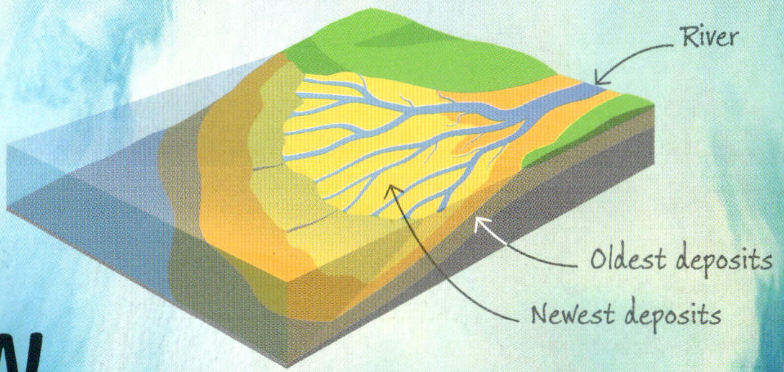

River

Oldest deposits

Newest deposits

▲ GANGES DELTA

The largest delta in the world, the Ganges Delta in India and Bangladesh is about three times larger than Belgium. Around 280 million people live there. In this satellite image, huge clouds of pale sediment are visible flowing into the Indian Ocean.

HOW DELTAS AND ESTUARIES WORK

Rivers carry vast amounts of sediment (sand and mud) from land to sea. The largest rivers dump so much sediment at the coast that it builds up to form new land – a delta. Smaller rivers end at wide, muddy mouths called estuaries. Here, the river and tide are in a constant battle to move sand and mud around.

ESTUARIES

In an estuary, mud and sand are pushed inland when the tide rises, but flushed back out by the river as the tide falls. These forces create large channels in between islands of sediment. Estuaries with stronger inward currents become clogged with mud and are bad for ships. Estuaries with stronger outward currents are more stable, with deeper channels.

HOW COASTLINES CHANGE

SHORE PLATFORMS

As a rocky coastline retreats over time, a horizontal or gently sloping rock surface can be left behind. This may have rock pools that get covered at high tide and exposed at low tide.

Coastlines are a constantly changing feature of our planet. The rocky cliffs of many coasts are continuously eroded by pounding waves, carving out caves and weakening cliff faces, which eventually crumble into the sea. Waves also deposit sand and rocks on the coastline, building up beaches that stretch along the shore.

Chalk is a soft sedimentary rock that erodes easily, forming white cliffs.

▶ ROCKY COASTLINES

The gradual wearing down of rocky coastlines creates amazing scenery, such as these chalk formations, known as Old Harry Rocks, off the Dorset coast in the UK. While the waves erode the cliffs, they also deposit rocks and sand in the bays along the coast.

This sea stack was once connected to the land.

BLOWHOLES
A blowhole forms when the roof of a sea cave collapses, leaving a hole in the top of the cliff. When waves gush into the cave, the water is squeezed upwards through the hole, creating a fountain of sea water.

SANDY COASTLINES
Beaches form from the build-up of pebbles and sand carried by waves. They constantly change shape. In calm conditions with gentle waves, sand and pebbles slowly pile up, making a beach steeper. In stormy weather, large waves wash sand and pebbles back out, making the beach flatter.

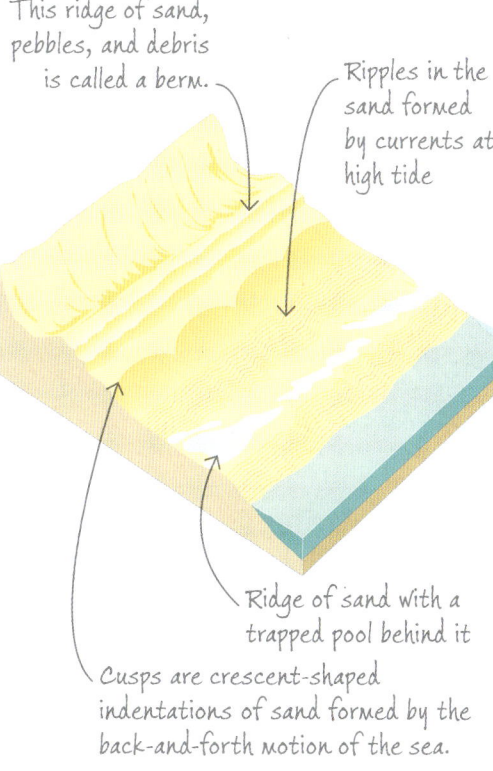

This ridge of sand, pebbles, and debris is called a berm.

Ripples in the sand formed by currents at high tide

Ridge of sand with a trapped pool behind it

Cusps are crescent-shaped indentations of sand formed by the back-and-forth motion of the sea.

LONGSHORE DRIFT
If waves wash up on a beach diagonally, they push sand and pebbles up the beach diagonally too. However, the water flowing back to the sea (backwash) pulls sand and pebbles straight back down. The overall result is a zigzag movement of sand and pebbles along the beach – a process known as longshore drift. Barriers called groynes are often placed on beaches to control erosion by longshore drift.

HOW A SEA STACK FORMS
The force of waves pounding against cliffs can open up cracks in the rock, which over time get bigger to form caves, and then bigger still to form arches. When an arch collapses it can leave a sea stack – a tall column of rock standing in the sea.

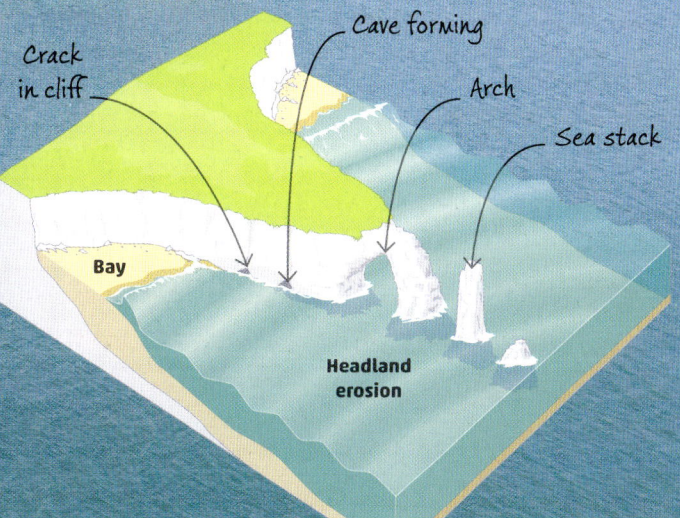

Crack in cliff

Cave forming

Arch

Sea stack

Bay

Headland erosion

INCOMING WAVES →   **BACKWASH** →

PREVAILING WIND

Groynes at Eastbourne, UK

Groyne

HOW
WAVES
WORK

Ocean waves form when the wind blows over the sea's surface. Waves absorb the wind's energy and can carry it for thousands of miles before breaking at beaches where there may be no wind at all. Although the water in waves appears to travel horizontally, it is actually moving up and down as energy passes through it.

German surfer Sebastian Steudtner holds the world record for the biggest wave ever surfed.

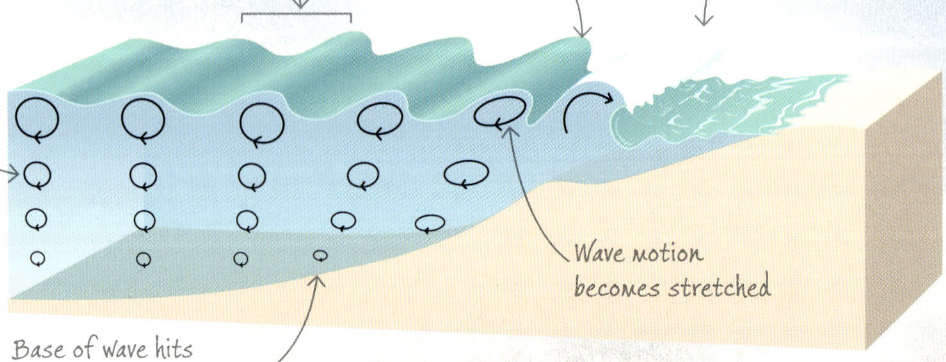

Wavelength is the distance from crest to crest or trough to trough

Crest begins to tip over

Wave breaks on the beach

Circular motion, which becomes smaller at depth

APPROACHING THE SHORE
Out at sea, the surface water forms a circular motion as each wave passes. As the wave approaches shallow water, its circular motions become more stretched, before it breaks on the beach.

Wave motion becomes stretched

Base of wave hits seabed and slows down

Rescuers on Jet Skis are ready if surfers fall under the waves at Nazaré.

FREAK WAVES

Occasionally, two large ocean waves combine to create a single giant wave called a freak wave. This can happen when two sets of waves collide at just the right angle or when a storm makes waves combine randomly. Freak waves taller than an eight-storey building can sink ships.

Rip current   Incoming wave

RIP CURRENTS

On beaches with an uneven floor, the water carried inland by waves may rush back out in a single powerful channel where the sea is deeper. This is called a rip current and can be deadly. Every year, hundreds of unwary swimmers are swept out to sea by rip currents and drown trying to swim back against the powerful flow.

◄ TOWERING WAVES

As waves approach land and the water becomes shallower, they slow down and grow taller. At Nazaré on the coast of Portugal, the waves are huge – as tall as 25 m (80 ft) – and attract surfers from all over the world. Energy from incoming surface waves here combines with wave energy from an underwater canyon to create the largest waves ever surfed.

The froth formed by breaking waves is known as spume.

BENDY WAVES

Just as light waves bend (refract) as they pass into glass and slow down, water waves can bend when they hit land. If one end of a wave reaches shallow water first, it slows down and the whole wave bends. Bending affects how waves erode land. Areas of land that stick out into the sea are hit by the fast ends of waves and are eroded, while slower ends of waves tend to deposit sand, forming beaches.

Look closely at a rock and you can sometimes see hundreds of tiny interlocking **crystals**. These are **minerals** – the solid, crystalline chemicals that are Earth's building blocks. Minerals crystallize into their solid form in various ways, such as when **molten rock** cools down and solidifies. They vary from tiny flecks of dull grey grit to glittering **gemstones** with brilliant colours.

ROCKS AND MINERALS

HOW THE ROCK CYCLE WORKS

Earth's rocky crust has been around for more than 4 billion years, but it's rare to find a rock this old. Rocks are constantly being broken down and recycled in a process known as the rock cycle. Each rock, stone, and pebble has a different story to tell.

► THE ROCK CYCLE

In the rock cycle, old rocks get transformed into new rocks by erosion, heat, pressure, or a combination of these things. For instance, deep in the Earth's crust, high temperatures and pressure changes melt ancient rocks to form magma, which rises towards the surface and solidifies as it cools. Most changes take hundreds, thousands, or even millions of years.

COMPACTION AND CEMENTATION

WEATHERING AND EROSION

SEDIMENTS

WEATHERING AND EROSION

WEATHERING AND EROSION

HEAT AND PRESSURE

By looking closely at a rock, you might be able to see grains or crystals of the minerals it is made from.

This granite rock contains crystals of several different minerals, including quartz, feldspar, and biotite.

MELTING

MELTING

Quartz is made from silicon and oxygen. Its chemical name is silicon dioxide, SiO_2.

COOLING

IGNEOUS ROCK

If rocks deep underground are heated enough, they melt to form magma. This is less dense than solid rock. It may rise into the upper part of Earth's crust or even erupt onto the surface. Here it cools to form igneous rocks, such as andesite and basalt.

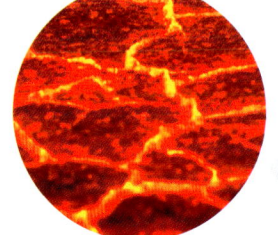

MAGMA

WHERE ROCKS FORM

Sedimentary rocks form on Earth's surface, as layers of sediment build up at the bottom of rivers, seas, and lakes. Metamorphic rocks form deep within Earth's crust, or in other areas of high heat and pressure. Igneous rock forms either underground when molten rock (magma) cools slowly, or on Earth's surface when lava spilled by a volcano cools quickly.

SEDIMENTARY ROCK

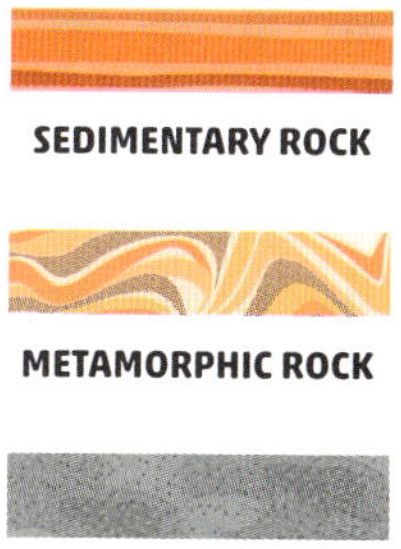

METAMORPHIC ROCK

IGNEOUS ROCK

SEDIMENTARY ROCK

Rocks on Earth's surface are broken down by weathering and erosion, forming mud and sand, which settle in layers at the bottom of rivers, lakes, and seas. Over time, the deepest layers are squeezed together and hardened by minerals that form in the tiny spaces between particles of sediment. This forms sedimentary rocks such as sandstone (above) and limestone.

HEAT AND PRESSURE

Weathering breaks down rock.

Rocks do not move through the rock cycle in a particular order. Any type of rock can change into any other type of rock.

SPACE ROCKS

Meteorites are rocks that fall to Earth from space. Most contain lots of iron or nickel, which makes them magnetic and heavy for their size. Many have a smooth black crust of rock that melted as the meteorite zoomed through our atmosphere.

MELTING

METAMORPHIC ROCK

As Earth's crust is changed by tectonic forces, rocks may become buried even deeper. Intense heat and pressure may cause minerals to change without melting completely, forming metamorphic rocks such as quartzite, marble (above), and slate.

HOBA METEORITE NEAR GROOTFONTEIN, NAMIBIA

HOW IGNEOUS ROCK WORKS

Water and weather constantly attack and erode rocks at Earth's surface, yet the crust is not getting thinner. New rock is added all the time as molten rock wells up, cools, and then hardens. Rocks formed by molten rock that cools and hardens is known as igneous rock, meaning "fire born".

Granite is a hard rock that takes a long time to weather and erode.

IGNEOUS GRAINS
Deep underground, magma cools slowly, and mineral crystals can keep growing for a long time before the rock sets solid. These coarse mineral grains are easy to see in granite.

PLAGIOCLASE FELDSPAR
Feldspars are the most common minerals in granite rock.

BIOTITE MICA AND AMPHIBOLE
Black grains are iron-rich minerals, such as biotite mica or amphibole.

POTASSIUM FELDSPAR
Pink granite gets its colour from potassium feldspar.

QUARTZ
At least a fifth of granite is made up of quartz.

MINERAL MOSAIC

Looking at a very thin slice of granite under a microscope reveals tiny interlocking mineral crystals. While most rocks are a mixture of two or more minerals, the minerals themselves are pure substances.

◀ GRANITE

Granite is one of the most common igneous rocks. It forms when molten magma cools deep beneath Earth's surface over many years. Granite's main mineral ingredients – quartz and feldspars – are both rich in silica (silicon dioxide). This is a clue that granite forms from shale or sandstone that has melted or partly melted deep underground.

CRYSTAL CLUES

Although all igneous rocks form from magma or lava, there is huge variation in the mineral crystals that make them up. The size of these grains gives igneous rocks different textures and tells you how quickly they cooled.

OBSIDIAN

Glassy obsidian forms when silica-rich lava cools so quickly (within hours) there is no time for mineral crystals to grow.

PUMICE

Pumice forms when the lava from a volcano mixes with water and gases. It cools and hardens quickly, so remains filled with tiny bubbles and rarely has visible crystals.

Blue topaz crystal in pegmatite

PEGMATITES

When magma cools slowly, crystals may keep growing for thousands or even millions of years. Pegmatites are rocks with crystals larger than 1 cm (0.4 in), including minerals that crystallize only at lower temperatures.

IN OR OUT?

Igneous rocks are classified as intrusive or extrusive, depending on where they form. Intrusive rocks, such as granite, form deep inside Earth's crust, often from magma that forced its way in between layers of existing rock before cooling slowly. Extrusive rocks form when magma exits the crust as lava and cools quickly.

LAVA COOLING TO FORM BASALT ROCK ON KĪLAUEA, HAWAII

HOW IGNEOUS INTRUSIONS FORM

Igneous intrusions are rock formations created by magma that melts its way through Earth's crust but fails to erupt from volcanoes. When erosion wears away the softer rock around intrusions, these masses of hard, resistant rock are left standing proud.

IN OR OUT
An igneous intrusion can take many different shapes. Runny magma may find its way horizontally between layers of rock, and more viscous (gloopy) magma rises more slowly towards the surface, cracking and melting the rock above it.

Stocks are similar to batholiths, but are smaller.

A volcanic plug forms when magma cools in a volcano's vent.

Laccoliths are mushroom-shaped intrusions.

Sills form in horizontal sheets parallel to existing rock layers.

Batholiths are huge and have irregular shapes.

Dykes are sheets that cut vertically through other rock layers.

A xenolith is a rock that gets trapped in the magma while it cools.

▼ **DEVILS TOWER (MATO TIPILA)**
This giant block of igneous rock in Wyoming, USA, is an igneous intrusion that formed deep underground about 50 million years ago, perhaps as a laccolith or a volcanic plug. Over time, the softer sedimentary rock surrounding it wore away, leaving the tough igneous rock standing like a giant monument.

The intrusion towers 264 M (866 feet) above the landscape.

ROCK COLUMNS

The igneous rock of Devils Tower forms spectacular hexagonal (six-sided) and pentagonal (five-sided) columns. Each one is hundreds of metres tall. They formed as molten magma cooled, shrank, and then cracked.

HALF DOME

Half Dome in Yosemite National Park, USA, is a batholith with the remains of an ancient magma chamber at its core. It originally formed as a dome but erosion has cut it in half, creating a sheer face on one side.

LA PALMA

Erosion can reveal the shape of a dyke within the surrounding rock, like this example on the island of La Palma in the Canary Islands. These frozen moments provide valuable clues about how lava finds its way through Earth's crust.

UNST

This laccolith on the island of Unst in Scotland got its distinctive dome shape when rising magma spread out horizontally between rock layers and pushed up the overlying layer of rock.

— Devils Tower consists of an igneous rock called phonolite porphyry.

FINGER MOUNTAIN

Finger Mountain in Antarctica gets its name from the vast dolerite sill that cuts through the layers of sandstone. Dolerite is a hard, dark rock that forms as magma cools.

Slate is usually grey, but can also be green, purple, or red.

2 SLATE
Close to Earth's surface where the temperature isn't very high, shale is transformed into slate. This low-grade metamorphic rock is brittle with fine grains.

3 PHYLLITE
Slightly higher temperatures and pressures can transform slate into phyllite. Its tiny, neatly arranged mica crystals reflect light, giving phyllite a glossy sheen.

1 SHALE
Shale is a sedimentary rock that is fine-grained due to being formed from compacted layers of clay and silt.

▶ GRADES OF METAMORPHIC ROCK
When rock such as shale is heated by superhot magma and squeezed with enough force, it changes into metamorphic rock. Some rocks go though multiple transformations, changing from one metamorphic rock type to another. The grade tells you how much the rock has changed – the higher the grade, the greater the transformation.

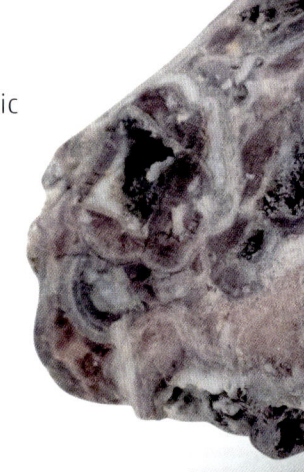

HOW
METAMORPHIC
ROCK
WORKS

Around 100 km (62 miles) beneath your feet, Earth's mantle becomes hot enough to melt rock into magma. However, rocks much nearer the surface can also find themselves facing extreme conditions. Along the faults where Earth's giant tectonic plates meet, and around magma that is forcing its way to the surface, existing rocks can be heated or squeezed so much that they change into new forms without melting completely. We call rocks that form this way metamorphic.

UNDER STRESS
Stripes, folds, or waves are often the first clue that you are looking at metamorphic rock. These patterns may be microscopic or enormous. Largest of all are the metamorphic rock formations that develop within mountain ranges as colliding tectonic plates squeeze and crumple Earth's crust.

Dark bands often contain heavy minerals, such as hornblende.

Pale bands are often rich in silica minerals, such as quartz and feldspars.

4 SCHIST
At even higher temperatures metamorphism goes further, transforming phyllite into medium-grade schist. The mineral grains in schist are large enough to see.

Wavy patterns developed in this schist as the minerals recrystallized during metamorphosis.

5 GNEISS
At very high temperatures and pressures, shale transforms into high-grade gneiss, which often has stripy patterns.

TRANSFORMATION
Metamorphic rocks may be altered slightly or transformed so much that it's hard to work out their original forms – rather like butterflies emerging from a caterpillar's cocoon. It's not just their appearance that changes either, as metamorphic rocks often have completely different properties from their parent rocks (protoliths).

SLATE
When shale is transformed by pressure into slate, certain minerals line up at right angles to the direction of the squeeze. This creates flat layers that flake apart easily, making slate a popular material for roof tiles.

MARBLE
Marble forms when limestone comes under heat and pressure. The mineral calcite recrystallizes to form interlocking crystals, making marble much tougher than limestone.

IMPACTITE
Not all metamorphic rock forms deep underground. The shock of a meteorite strike can change rock, sand, or soil into rocks called impactites. These include rare types of natural glass that were prized as gemstones in ancient Egypt.

DIFFERENT TYPES OF SEDIMENT

The particles that sedimentary rock forms from – and the way it forms – determine whether the rock will be soft or hard. Many sedimentary rocks contain fossils, and some types consist almost entirely of fossils.

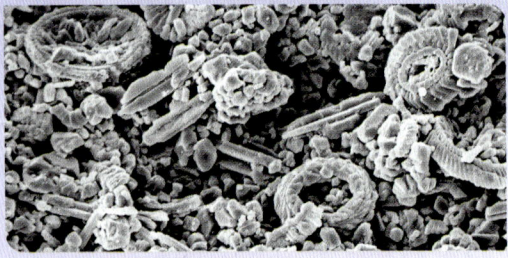

LIMESTONE AND CHALK

Chalk and many types of limestone are made of the fossils of marine organisms. This microscope image of chalk shows fragments of tiny shells made of calcium carbonate.

SANDSTONE

About a fifth of all sedimentary rock is sandstone, which is made of ancient sand grains. Sandstone is often used as a building material as it is durable but easy to carve.

MUDSTONE AND SHALE

Mudstone and shale form when very fine particles of mud and clay are tightly compressed, leaving little room for mineral cement between the grains. This makes these rocks soft and flaky.

CONGLOMERATE

Conglomerate is made from larger bits of rock from grit and gravel to pebbles and even boulders dumped by fast-flowing water. These are all held together by a finer-grained material called a matrix.

HOW SEDIMENTARY ROCK WORKS

At Earth's surface, rocks of all kinds are gradually worn away. Over millions of years, weathering and erosion even break mountains into tiny particles of silt, sand, and mud. These are washed into rivers and oceans where they settle to the bottom. Over millions more years, the sediments are squeezed and cemented together as minerals form in the tiny spaces between them. They become sedimentary rock.

Rivers wash sediment from land into the sea.

DEPOSITION

COMPACTION

CEMENTATION

Over time, sedimentary rock may be lifted and tilted by the movements of Earth's tectonic plates.

FORMATION OF SEDIMENTARY ROCKS

Sedimentary rock tends to form flat layers called strata. As new sediment settles at the top (deposition), its weight squeezes lower layers (compaction). Water seeping through the strata carries dissolved minerals that crystallize between the particles and bind them (cementation). Sedimentary rock can also form chemically when minerals dissolved in water crystallize into solid form.

▼ LAYER UPON LAYER

Weathering in Argentina's Gorge of the Shells has revealed spectacular layers of sandstone, siltstone, shale, and conglomerate. These layers formed from sediments that were carried by fast-flowing water. As the water emerged from a small gap it fanned out, slowed, and the sediments settled to the bottom.

Most of the layers are sandstone and siltstone. The red and orange colours come from iron that has oxidized (rusted) in the air.

MICROSCOPE VIEW

If you look at sandstone with a microscope, you can see the individual grains along with the matrix that cements them together. Other types of sedimentary rock, such as mudstone, may also contain microscopic fossils such as pollen grains, which help reveal the rock's age.

Some of the layers are conglomerate, a type of sedimentary rock with large grains.

The green-grey layers are shale.

HOW SOIL WORKS

Soil is the skin that covers Earth's rocky crust. It's made up of particles of rocks and minerals from the crust, mixed with air, water, decaying plants and animals, and countless living organisms too. The soil layer makes up a tiny fraction of the planet, but it provides water and essential nutrients to plants as well as a home to many animals.

SAND, SILT, AND CLAY

Around half of soil is made up of rock particles. The three main types are sand, silt, and clay. Sand particles are largest and help water flow, so sandy soils tend to be drier. Clay particles are smallest and hold on to water, so clay-rich soils can be wet and sticky. They also retain nutrients well and so are more fertile than sandy soil.

Sand particles help water drain away easily.

FREE DRAINING

Silt particles are smaller than sand but bigger than clay.

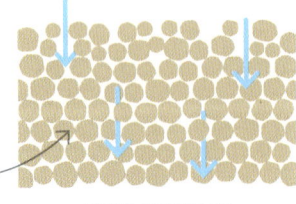

GOOD DRAINAGE

Water clings to tiny clay particles, helping soil retain moisture.

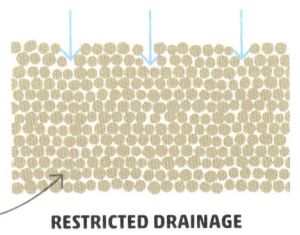

RESTRICTED DRAINAGE

Rich in organic matter, the upper layer of topsoil tends to be darker than the layers below.

As plant roots push down through soil, they create spaces for water and air. As roots grow and expand, they can crack and break down rock.

Soluble substances such as calcium, sodium, and potassium dissolve in water and then leach (drain) away and accumulate in lower layers.

Below the soil is the bedrock of Earth's crust. Moisture and biological activity in soil slowly weather this rock, and break it down into small particles.

BACTERIA

Soil contains countless trillions of microscopic bacteria, which are vital for its health. Among other things, they play a crucial role in nitrogen fixation. This involves taking nitrogen from the atmosphere and converting it into nitrates, which plants need in order to grow.

WORMS

Worms tunnel through soil, eating decaying organic matter, fungi, and bacteria. This helps to recycle the nutrients in soil so that they can be used by plants. The tunnels also help oxygen enter soil and reach the roots of plants.

SPRINGTAILS

Tiny insect-like animals known as springtails live in the top layer of healthy soils. They feed on organic matter, which helps to break down and recycle nutrients. Each square metre (10 sq ft) of soil contains an estimated 100,000 springtails, which makes these tiny creatures the most numerous and successful animals on Earth.

◄ SOIL HORIZONS

Soils have distinct layers known as horizons. The upper layer of topsoil, called humus ("hyoo-muss"), is rich in dead organic matter. It's also rich in tiny living animals, fungi, and microorganisms. These all help to break down the decaying matter and release nutrients, which plants absorb and recycle. Further down, soil becomes richer in rock particles that are slowly weathering.

SOILS IN DIFFERENT CLIMATES

Soils form gradually, beginning with sediments that have been eroded from a mountain, deposited by water, or carried on the wind. Over time, the local climate, living creatures, and the landscape all play their part in shaping a soil's story.

FROZEN SOIL

Near the poles, and at high altitudes, the water held by soil can stay frozen for years at a time. This "permafrost" pauses the process of decay, which means the frozen soil can store carbon that would otherwise be released as greenhouse gases.

DESERT SOIL

Soil forms slowly in deserts. The dry climate means there is less weathering of the underlying rock, and fewer living things to become humus when they die. However, low rainfall also means nutrients in desert soils are less likely to be leached away. They often collect as a rock-hard, white layer below the surface.

RAINFOREST SOIL

Rainforests have only a thin layer of topsoil and are low in nutrients. High temperatures allow microbes to digest decaying matter in hours, and dissolved nutrients are soon leached away by heavy rain. Only iron and aluminium oxides hang around, giving rainforest soils the red colour of rust.

HOW **MINERALS** WORK

Crystals inside a geode grow inwards towards the centre.

A dazzling display of minerals can rival Earth's most colourful plants and animals. Minerals are the chemical building blocks of rocks. They grow naturally – often deep underground – but they are not alive. A mineral is simply a solid substance made of a specific set of chemical ingredients. Every mineral has a unique set of properties, which come in handy for working out which of Earth's 6,000 or so named minerals you're looking at.

▶ NATURAL TREASURES

Most rocks are made of minerals, but geodes are famous for putting their building blocks on spectacular display. Geodes are roundish, ordinary-looking rocks from the outside, but breaking one open reveals a hollow cavity lined with mineral crystals. These crystals grow from minerals deposited by water and can take thousands of years to form.

Many geodes are filled with visible crystals of quartz, one of the most common minerals found on land.

AGATE CRYSTALS

Geodes often contain a special form of quartz called agate. Its crystals are too small to see with the naked eye, but they are visible through a microscope. They are coloured by tiny traces of other substances and build up in layers, forming wavy patterns called arcs.

NEAT AND ORDERED
Minerals are pure substances, so their chemical ingredients can arrange themselves in regular, repeating 3D patterns. These orderly insides give each mineral's crystals a precise geometric shape, which can help identify it.

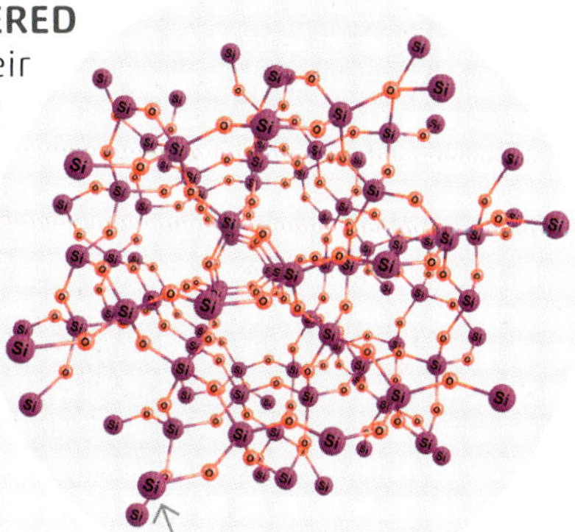

Atoms of silicon and oxygen in a quartz crystal

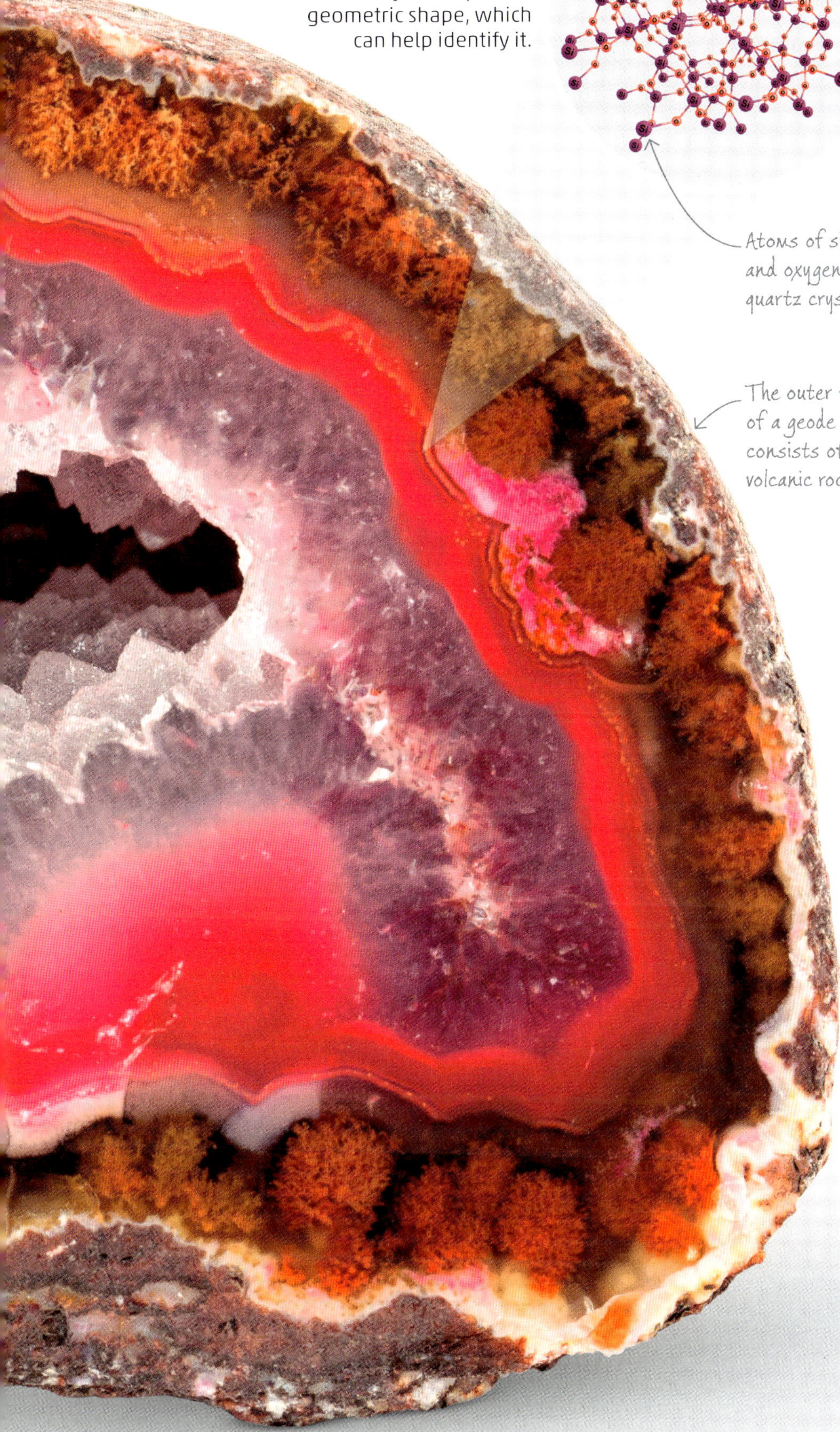

The outer part of a geode often consists of volcanic rock.

CLASSIFYING MINERALS
Scientists group and name minerals based on their main chemical ingredients. For example, silicate minerals contain the elements silicon and oxygen, often with metal elements too.

Carbonates
Copper gives malachite its green colour, but it's the non-metal elements (carbon and oxygen) that make it a carbonate.

Silicates
More than a thousand different minerals are classed as silicates, making this the largest group of minerals.

USING MINERALS
Minerals have shaped human history from the Stone Age, when people learned to strike flint with pyrite to create sparks and start fires. Since then, we have found millions of ways to put different minerals to use, including more than 42 minerals used to manufacture a typical smartphone.

MINING MINERALS
Rocks that contain useful minerals are known as ores. We get them from mines with digging machines, though doing so uses lots of energy and can harm the natural environment.

HOW CRYSTALS WORK

Most pure solid substances, from minerals to metals, are made up of crystals. These are often too small to see, but in some rocks they are visible as separate grains or as larger geometric shapes. Crystals form when the atoms or molecules in a substance become arranged in an orderly, repeating pattern. This can happen when the minerals in molten rock (magma) cool and solidify, or when the water around dissolved minerals evaporates.

▼ **PYRITE**

The shape of a crystal depends on the structure of its atoms and molecules. In the mineral pyrite, iron and sulphur atoms are arranged in repeating units that are cube-shaped, and so the crystals are cube-shaped too. Pyrite is also called fool's gold because it's sometimes mistaken for gold. A good test is to scratch it with a fingernail – this would leave a mark on gold, but not pyrite.

The cubes of pyrite are natural, despite their appearance.

ATOMIC CUBES

The iron and sulphur atoms inside pyrite are arranged in repeating cubes, giving pyrite crystals their seemingly unnatural shape.

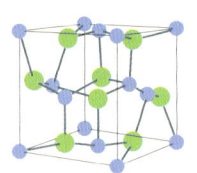

Many repeating cubic structures

CRYSTAL SYSTEMS

A crystal begins as a single unit (a group of atoms or molecules). As the crystal grows, more units are added. In perfect conditions, they build up neatly in three dimensions along axes (imaginary straight lines). The shape of a crystal depends on the number of axes, their length, and the angle between them, but all crystals can be grouped into one of seven basic classes called crystal systems.

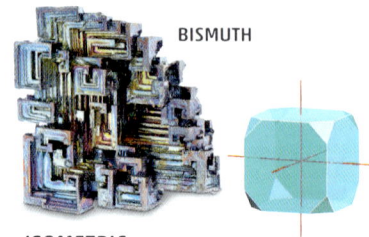

BISMUTH

ISOMETRIC
Three axes of equal lengths are at right angles to each other. Isometric crystals include cubes, octahedrons, or dodecahedrons.

VESUVIANITE

TETRAGONAL
Three axes are at right angles to each other, but one is longer than the other two. Tetragonal crystals include cuboids or square blocks with a point at each end.

TOPAZ

ORTHORHOMBIC
Three axes, all with different lengths, are at right angles to each other. Orthorhombic crystals can look like tetragonal crystals that have been squashed in one direction.

ALLOTROPES

The exact same atoms can make different crystals in different conditions. At high pressure, carbon atoms form hard diamond crystals with atoms arranged in tetrahedral (pyramid) structures. At low pressure, they form flat layers of soft graphite, with weak bonding between layers. These two forms of carbon are known as allotropes of carbon.

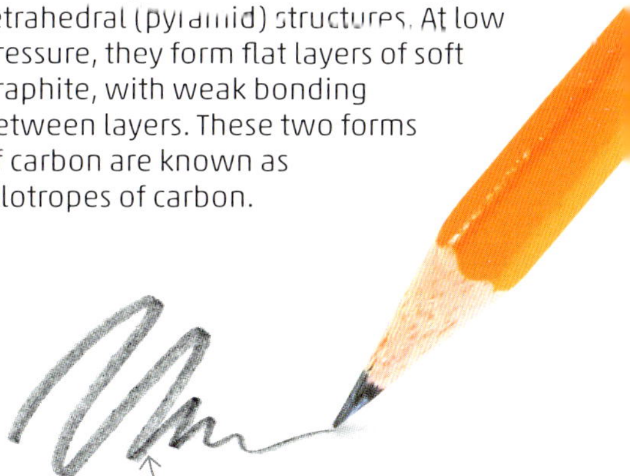

Graphite is soft. It's the substance inside pencils.

Diamond is the hardest mineral on Earth.

GIANT CRYSTALS

In the right conditions and given enough time, crystals can grow to gargantuan sizes, like these crystals of gypsum discovered in the Naica Mine, Mexico.

EMERALD

HEXAGONAL
Three horizontal axes are at 120° to each other, plus one vertical axis. Hexagonal crystals can look like six-sided blocks with a point at each end.

AMETHYST

TRIGONAL
Three horizontal axes are at 120° to each other, plus one vertical axis. This is like hexagonal, except there are fewer lines of symmetry. Trigonal crystals can look like a triangular block with a point at each end.

KUNZITE

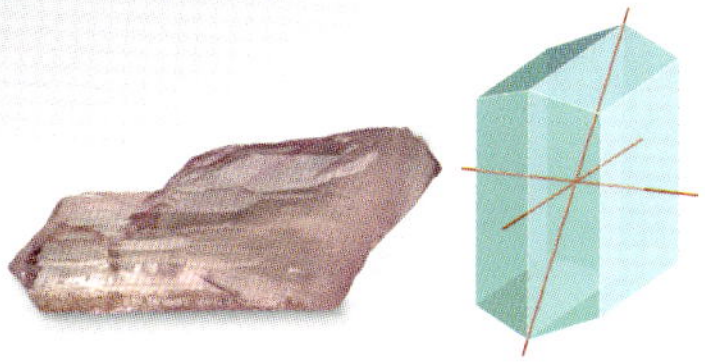

MONOCLINIC
Two axes are at right angles to each other, and a third axis is on a tilt. All three axes can be different lengths. Monoclinic crystals have end faces that are tilted to one side.

AMAZONITE

TRICLINIC / RHOMBOHEDRON
Three axes are at different angles and have different lengths. Triclinic crystals are the least symmetrical and can often have a seemingly random shape, despite following a system.

CRYSTAL HABITS

The microscopic building blocks of a mineral's crystals are neat and orderly geometric shapes. However, when crystals build up in nature they can grow into more complex shapes, from fibres and needles to grapelike clusters and flat sheets. These distinctive shapes have special names and are known as crystal habits.

FIBROUS
Silicate minerals can form crystals so long, thin, and flexible that they look more like plant or animal fibres. Asbestos is the most infamous – if breathed in, asbestos fibres can damage people's lungs.

Fine fibres

ACICULAR
Crystals of mesolite grow into long and slender needles, like pins in a tiny pincushion. This habit, called acicular, makes the brittle crystals very fragile.

RADIATING
Wavellite crystals radiate outwards from a single point, forming spheres. When cut in half, the tightly packed, fibrous crystals look like sparkling starbursts.

CONCENTRIC
Malachite is usually found with a banded pattern of dark and light green layers, which sometimes grow out from a centre and look like circles. Cutting and polishing can reveal these patterns.

Clusters of tiny quartz crystals

BOTRYOIDAL
The name for these clusters of rounded masses means "bunch of grapes". Up close, each grape turns out to be a cluster of tiny crystals with flat geometric faces.

LENTICULAR
"Desert roses" form in dry, sandy places in Mexico when water evaporates and leaves dissolved gypsum behind. The lentil-shaped crystals form clusters like petals in a rose.

LAMELLAR
Lamellar means plate-like and describes minerals that form thin, flaky sheets. Mica forms the thinnest sheets of all – about the same thickness as a piece of paper.

AGGREGATE
Two or more types of crystal growing together are described as an aggregate. Here, globular groups of calcite seem to bulge out of a forest of quartz crystals.

TABULAR
Anhydrite crystals typically grow in flat tabular (table-like) shapes. They sometimes look like small notepads or playing cards clustered together.

Opposite faces remain parallel to one another.

PRISMATIC
When crystals grow slowly with plenty of space, they get the chance to form large, prismatic shapes that show off the inner geometry of their building blocks.

Beryl often forms pencil-like hexagonal crystals.

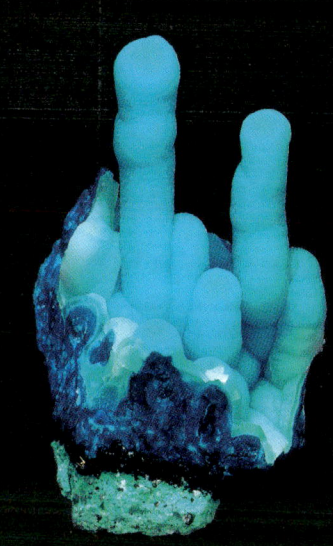

STALAGMITIC
Clusters of chrysocolla crystals can look like stalagmites but are found inside hollow rocks rather than inside caves. The beautiful blue-green colour comes from the element copper.

SCEPTRED
Crystals of "dog-tooth" calcite have sharp points. Sometimes the tips are separate crystals growing on the ends of older ones. This arrangement is called a sceptred habit.

WHY MINERALS GLOW

Never dismiss a dull-looking rock. It might have an extraordinary property known as fluorescence, which makes it light up in glowing colours when lit with an ultraviolet light. The most spectacular glowing rocks contain hundreds of fluorescent grains that sparkle like tiny individual lights, as though lit from deep within.

The cerussite crystals are brownish-white in normal light.

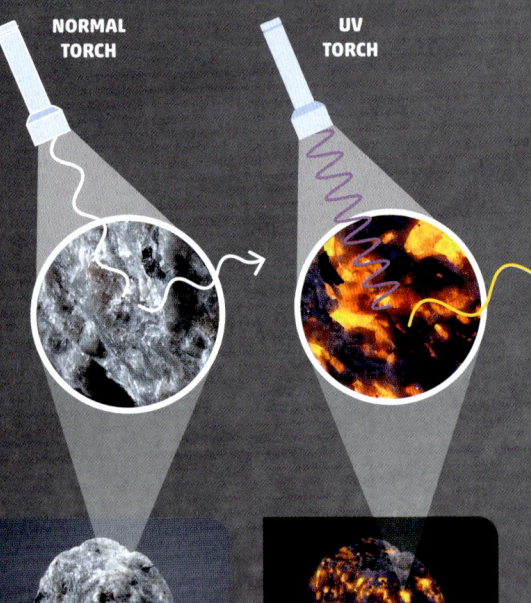

NORMAL TORCH

UV TORCH

SYENITE IN WHITE LIGHT

SYENITE IN UV LIGHT

HOW FLUORESCENCE WORKS
An ultraviolet (UV) torch is needed to see fluorescent minerals glow. UV light has a shorter wavelength but higher energy than normal light and is invisible to our eyes. When it hits a fluorescent mineral, such as this syenite, electrons in certain atoms absorb energy and jump to higher orbits in their atoms. As they fall back to their previous orbits, they emit light of lower energy and longer wavelengths – visible light.

▼ GLOW IN THE DARK

This rock contains crystals of the mineral cerussite (lead carbonate). Some cerussite crystals fluoresce with an intense yellow when lit by ultraviolet (UV) light. Geologists aren't sure if this is due to the mineral's high lead content or the presence of impurities such as silver. Fluorescent minerals stop glowing as soon as the UV torch is turned off, but other kinds of mineral have a more persistent glow called phosphorescence.

Cerussite glows bright yellow when lit by UV.

Minerals that aren't fluorescent look dark in UV light.

CALCITE
Calcite is one of the most common fluorescent minerals. Traces of the element manganese make it glow pink-orange.

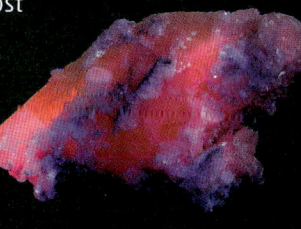

SPHALERITE
Sphalerite usually glows orange under UV light, but certain specimens light up in a rainbow of colours thanks to a mixture of impurities.

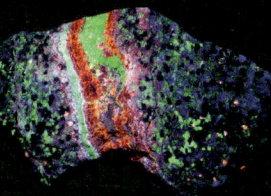

FLUORITE
Fluorescence was named after fluorite, the first mineral discovered to glow under UV light. Not all specimens glow – only those that contain the elements yttrium, europium, or samarium.

WILLEMITE
The green crystals in this rock are willemite, a mineral rich in zinc. It's one of the brightest phosphorescent minerals, with a glow that lasts a long time after UV light is switched off.

CORUNDUM
Rubies and sapphires are forms of the mineral corundum. Rubies glow bright red in UV light, but green and blue sapphires don't fluoresce.

SODALITE
Some syenite rocks are rich in the fluorescent mineral sodalite. When lit by UV, the sodalite crystals light up with a fiery orange glow.

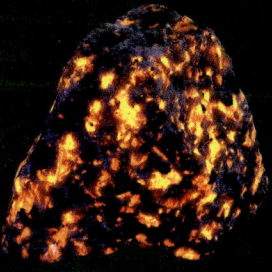

HOW PIGMENTS WORK

Rocks and minerals are among the most important sources of natural pigments – the colourful compounds used to make paints. Cave paintings made by people in the Stone Age show that people have been crushing and heating rocks to make colours for perhaps 100,000 years.

RED FOR DANGER

Cinnabar (mercury sulphide) is a mineral found near volcanoes and hot springs. It was once ground up to make a blood-red pigment called vermilion. This no longer happens because heating cinnabar releases dangerous mercury vapour.

POWDERED CINNABAR

CINNABAR

VERMILION PAINT

HAEMATITE

POWDERED HAEMATITE

RED OCHRE

IRON RED

Rocks and minerals containing iron oxides have been used to make pigments for tens of thousands of years. The earthy colours of yellow ochre, red ochre, and brown umber resemble the rocks and soils they're made from. They can also be roasted to make darker pigments.

POWDERED LAZURITE

LAZURITE

ULTRAMARINE PAINT

PRECIOUS BLUE

There are very few blue pigments in nature, so the mineral lazurite was once as precious as gold. It can be polished to make the gemstone lapis lazuli or crushed to create the brilliant blue pigment ultramarine.

PSYCHEDELIC GREEN

Malachite is a copper carbonate mineral and one of the oldest sources of green pigment. It's found in paintings on the walls of Ancient Egyptian tombs.

MALACHITE

POWDERED ORPIMENT

ORPIMENT

KING'S YELLOW PAINT

LETHAL GOLD

The mineral orpiment has a golden sheen that persists even when it's ground and turned into paint. Also called "king's yellow", it was used to decorate medieval manuscripts. There is no gold in orpiment. Instead it contains the deadly element arsenic and was once used to make lethal poisons.

CLASSICAL BLUE

Ancient Egyptian artists used a copper carbonate mineral called azurite to make blue pigments. The more finely it's ground, the lighter blue the powder becomes.

AZURITE BLUE PAINT

POWDERED AZURITE

AZURITE

POWDERED REALGAR

ARSENIC ORANGE PAINT

REALGAR

EGYPTIAN ORANGE

The mineral realgar grinds into a bright orange pigment, which was used to decorate papyri and tombs in Ancient Egypt. It's rarely used today because, like orpiment, it contains the toxic element arsenic.

CHALK

POWDERED CHALK

CHALK WHITE PAINT

WHITE PIGMENTS

Soft chalk was one of the first minerals to be ground and used as a white pigment. It's found in prehistoric cave art and is still a popular art material today.

POWDERED MALACHITE

MALACHITE GREEN PAINT

BLACK PIGMENTS

Graphite (a form of carbon) and manganese oxide minerals such as pyrolusite are both sources of black and dark brown pigments.

GRAPHITE

POWDERED GRAPHITE

BLACK PAINT

STREAK TEST

Many minerals vary in colour, which can make them tricky to identify. A streak test can be used to find their true colour. This is done by dragging the mineral along a white porcelain plate. Minerals harder than porcelain may need to be filed or crushed first.

ORPIMENT

HAEMATITE

CROCOITE

CHALCOPYRITE

CINNABAR

MOLYBDENITE

EMERALD

Emerald is one of the different varieties of the mineral beryl. Its deep green colour comes from an impurity – the element chromium. Emeralds are rare, which makes them precious. Other types of beryl, such as the pale blue mineral aquamarine, are more common and less expensive.

The symmetrical faces cut into gemstones are called facets.

EMERALD

UNCUT EMERALD IN ROCK

SAPPHIRE

Sapphires are made of a mineral called corundum (aluminium oxide). Their varying colours – from blue and green to yellow and hot pink – come from impurities. Blue sapphires contain traces of iron and titanium in place of some of the aluminium atoms. This allows the crystal to absorb all colours except blue, which is reflected.

BLUE SAPPHIRE

UNCUT SAPPHIRE IN ROCK

Cutting a gemstone makes light reflect inside it, enhancing its colour, sparkle, and value.

RUBY

Rubies are corundum crystals in which up to 1 per cent of the aluminium atoms are replaced by chromium. This impurity gives rubies their intense red colour. Like sapphires, rubies are very hard and durable, which is ideal for jewellery.

UNCUT RUBY IN ROCK

CUT RUBY

TYPES OF GEMSTONE

Once you learn the secrets of Earth science, every rock feels like treasure. But the rocks that are most treasured of all are gemstones. These crystalline minerals are valued for their sparkling colours. They are usually cut and polished to improve their shape and reflect light, which enhances their beauty.

JADE

The minerals jadeite and nephrite are both sources of the semi-precious gemstone jade, valued for its green colour and smooth texture. It is hard but can be carved into intricate shapes.

Gemstones can be polished to smooth pebbles by tumbling them in grit.

POLISHED JADE

UNPOLISHED JADE

GARNET

Garnets are made of silicate minerals. Like sapphires and rubies, they get their colour from impurities. Silicates are not rare, so garnets are not as precious as sapphires and rubies.

CUT GARNET

GARNETS IN ROCK

LAPIS LAZULI

Most gems are crystals of a single mineral, but lapis lazuli is a mixture of several minerals, including blue lazurite, white calcite, golden pyrite, and blue sodalite. This rare rock is prized for its intense blue colour and is polished into rounded gemstones (cabochons).

POLISHED LAPIS LAZULI

ROUGH LAPIS LAZULI

SUNSTONE

Feldspars are the most common minerals in Earth's crust, but they sometimes grow into large crystals that are used as gemstones. Sunstone is golden with a metallic sparkle caused by flecks of other minerals embedded in the feldspar.

ROUGH SUNSTONE

POLISHED SUNSTONE

MOONSTONE

Moonstone is a feldspar gemstone. It has a pearly colour with blue highlights that are iridescent. This means the colour shimmers and changes when viewed from different angles, like colours on a soap bubble.

ROUGH MOONSTONE

POLISHED MOONSTONE

OPAL

Opal is made of silica with an added ingredient: water. This makes it a mineraloid rather than a mineral. Opal does not have the ordered inner structure of crystalline minerals. Instead, each opal is a unique substance. The most precious opals have brilliant iridescence, displaying a rainbow of different colours. Black opals are rarer than diamonds and can be more expensive.

RAW OPAL

POLISHED OPAL

HOW **DIAMONDS** WORK

Impurities in the diamond crystal or exposure to natural radiation produce green diamonds.

Every year billions of pounds are traded for tiny pieces of diamond, a mineral made of carbon – the same element as in coal and the graphite in pencils. What makes diamonds so special? For jewellers, it's their rarity and their brilliant sparkle. For toolmakers, it's their extreme hardness. For Earth scientists, it's the story of their explosive journey to Earth's surface.

Tetrahedral shape

INSIDE A DIAMOND
Diamond's hardness comes from the way the atoms bond when carbon crystallizes at high temperatures and pressures. Each atom forms strong bonds with four other carbon atoms in a tetrahedral (triangular pyramid) shape. This shape resists compression in every direction, making diamond very strong.

The mineral olivine gives kimberlite its greenish colour.

Diamond embedded in kimberlite

DIAMOND PASSENGERS
All natural diamonds are very old – a billion years or more – and formed at least 150 km (90 miles) underground. Geologists are still unsure exactly how diamonds form. Most found in mines are embedded in a rock called kimberlite. This rock forms from magma that rises to Earth's surface during volcanic eruptions. On its journey upwards, kimberlite carries gemstone passengers, including diamonds.

Bort diamonds are low-quality grainy crystals, usually crushed and used as abrasives in industry.

▼ ROUGH DIAMONDS

Natural, uncut diamonds come in a range of shapes and colours. Some crystals are octahedral – like two pyramids joined at their square bases – and others are round or irregular in shape. The most common colours are white, yellow, and brown, but diamonds can also be green, orange, red, and blue.

The quality of this octahedral crystal makes it suitable for being cut into a gemstone.

DIAMOND DRILL

Diamond is the hardest natural substance, which makes it ideal to cut or drill into other hard substances, such as rocks, metals, and gemstones, including other diamonds. The diamond-tipped drill bit shown here is polishing metal.

CUTTING DIAMONDS

Like other gemstones, diamonds are cut into a range of symmetrical shapes with lots of facets (flat faces). The facets reflect light inside the jewel, making it sparkle. Diamonds are so hard that they can only be cut and polished using tools coated in diamond.

  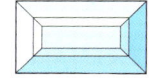 

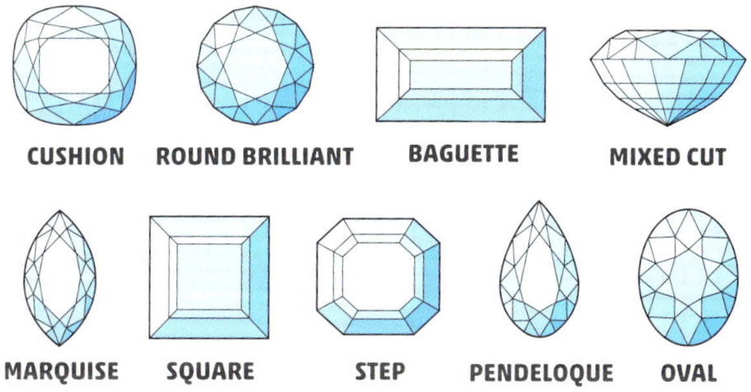

CUSHION   **ROUND BRILLIANT**   **BAGUETTE**   **MIXED CUT**

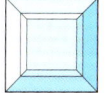 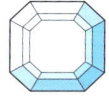

MARQUISE   **SQUARE**   **STEP**   **PENDELOQUE**   **OVAL**

DIAMOND FIRE

Diamond refracts (bends) light more powerfully than glass. As a result, light bounces around inside a cut diamond and splits into a rainbow of colours as it leaves. This gives diamonds a colourful sparkle known as diamond fire.

BRILLIANT-CUT DIAMOND

HOW NATIVE ELEMENTS WORK

The chemical elements are the building blocks of matter, each consisting of a unique kind of atom. Most of the matter around us consists of compounds – substances in which atoms of different elements are bonded together. However, a small number of elements can be found in their pure form in Earth's crust. These are known as the native elements.

The vertical columns are called groups.

The horizontal rows are called periods.

PERIODIC TABLE
The periodic table is a chart that lists all the elements in order of their atomic number (the number of protons in an atom's nucleus). Elements that fall in the same column have similar chemical properties. Every element has a unique chemical symbol. For instance, carbon is C, gold is Au, silver is Ag, and copper is Cu.

▼ NATIVE SILVER
Silver is classed as a "precious metal" because it is relatively rare. It has long been used as a form of money and in ancient times was more highly valued than gold. Today both metals are important in electronics as they are good at conducting electricity. Silver is also "ductile", which means it can be drawn into wires. These can be so fine that a single gram of silver can stretch 2 km (1.2 miles).

Silver tarnishes (loses its shine) because it reacts with sulphur in the air to form silver sulphide. This coating can be polished off to make silver bright again.

When native silver crystallizes in hollows, it forms amazingly varied shapes, from flakes and plates to finely branched crystals and long, curly wires.

Native silver can have a filiform (wirelike) habit.

GOLD

As gold doesn't react easily with other elements, it doesn't tarnish and keeps its shine. Along with Its rarity and colour, this makes it one of the most precious metals.

PLATINUM

Rarer even than gold, platinum is more valuable. It also has a higher melting temperature and so platinum ornaments were traditionally hammered into shape.

COPPER

The world's third most-used metal, copper is used to make electric wires. Around 5,000 years ago, people found they could make a stronger metal by mixing it with tin, which triggered the Bronze Age.

SULPHUR

Crystals of pure, bright yellow sulphur can be found in rocks near volcanoes. Although it is famous for forming stinky compounds, pure sulphur has no smell.

CARBON

In the forms of coal, diamond, and graphite, carbon has been used for thousands of years. Only in the 1700s did chemists realize that these are all different forms of the same element.

ORES

Iron and many other useful metals are found in rocks as chemical compounds rather than native elements. These metals are extracted from their ores (rocks) by melting, roasting, or chemical reactions.

SULPHUR PONDS

One of the most alien places on Earth is the Danakil Depression in Ethiopia. Here, hot springs stained yellow and orange by the elements sulphur and iron sit among active volcanic vents belching the stench of rotten eggs. The magma-heated spring water is close to boiling, saturated with salts, and incredibly acidic, yet strange forms of microorganisms have been found thriving in this scalding chemical soup.

BIOMINERALS

While most minerals are found as rocks, a small number can be found inside living creatures. These biominerals are created by their hosts for support, defence, as sensors, or to store useful substances. More than 60 different biominerals have been found, in microbes, plants, animals, and fungi. It's even thought that mineral crystals played a starring role in getting life started in the first place.

PLANT PROTECTION
Crystals of calcium oxalate are found in plants of all shapes and sizes. Needle-shaped crystals form in leaves and stems to deter plant-eating insects, by damaging their mouthparts. Some plants have stinging spines or hairs tipped with calcium oxalate to defend against grazing herbivores.

VIRGINIA CREEPER

CALCIUM OXALATE NEEDLES

JEWELS OF THE SEA
The surface waters of the sea teem with trillions of tiny, plantlike microorganisms called phytoplankton. Many of these single-celled organisms protect themselves with shells made of the minerals silica or calcium carbonate. Diatoms have such ornate silica shells that they are sometimes called the jewels of the sea.

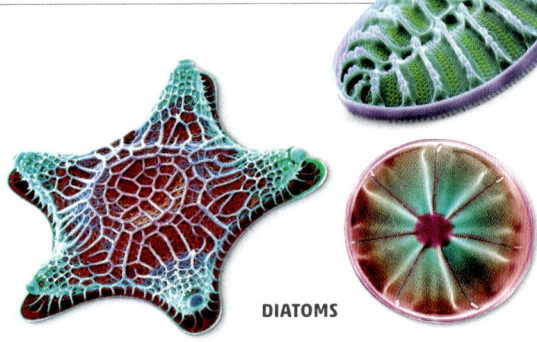

DIATOMS

The calcium carbonate shell is left behind after the mollusc dies and decays.

CHALK
When phytoplankton die, their remains sink to the sea floor. Over millions of years, deep layers of phytoplankton shells can build up and turn into rock. Chalk is a rock formed from broken bits of calcium carbonate that once formed the shells of organisms called coccolithophores.

COCCOLITHOPHORE

CHALK CLIFFS, UK

Nacre's iridescent shimmer is prized by makers of jewellery, furniture, and musical instruments.

PEARLS
As well as coating their shells with nacre, oysters and mussels sometimes cover flecks of dirt with the mineral too. After several months or years, the nacre can build up to form a pearl. These shiny blobs of waste have been used by humans as jewellery for centuries.

MOTHER-OF-PEARL
Some molluscs toughen their shells with a lining of nacre (mother-of-pearl). This shimmering mineral is a special form of calcium carbonate, also known as aragonite. It is laid down in tiny plates, sandwiched with stretchy protein to prevent cracks. It makes the shell very resilient to crabs or fish.

PEARL IN PACIFIC OYSTER

SPIRAL SHELL
Many molluscs, from land-living snails to their marine relatives, build shells from aragonite to protect their soft bodies from predators and from drying out. The mineral is laid down in layers and extended as the mollusc grows. This gives some seashells and snail shells their spiral pattern.

SUITS OF ARMOUR

A crab's exoskeleton begins as a bendy scaffold made of protein and carbohydrate. As gaps in the scaffold are gradually filled with the mineral calcite (a type of calcium carbonate), the shell hardens into formidable armour.

REEF BUILDERS

Coral reefs are built by tiny, jellyfish-like creatures called polyps. Each one builds a tough calcium carbonate shelter, popping out its tentacles to gather food. When polyps die, a new generation of skeletons forms on top of the old one. In this way, coral reefs slowly grow over millions of years.

CORAL SKELETON

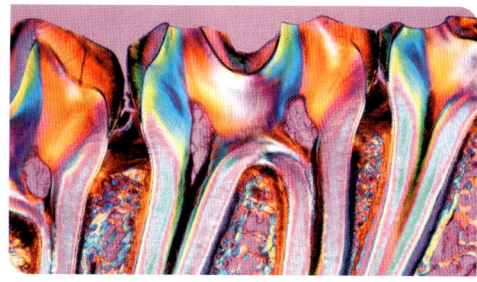

BONES AND TEETH

The building blocks of human bones and teeth are tiny crystals of apatite, a calcium phosphate mineral also found in rocks. Running throughout every bone is a network of living cells that control how apatite crystals are arranged. The enamel on teeth is 95 per cent apatite, making it the toughest substance in the body.

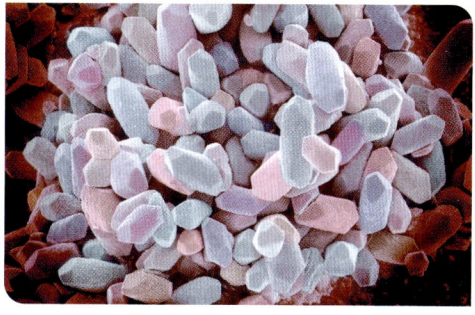

IN BALANCE

Otoliths (ear stones) are tiny crystals of calcium carbonate that form deep inside your inner ears. Each time your head moves, the crystals move a tiny amount in their bed of gel. This sends signals to your brain, which tells your muscles how to adjust your position so you stay balanced.

UNWANTED MINERALS

Certain illnesses are caused by mineral crystals growing in the wrong place in a person's body. For example, kidney stones grow from dissolved salts in urine. This microscope image shows the microscopic crystals of calcium oxalate sticking up from the surface of a kidney stone.

NAUTILUS SHELL

Life on Earth would not exist if our planet did not have an **atmosphere**. This thin blanket of **gases** protects us from the fierce glare of the Sun and the freezing vacuum of outer space. It gives us breathable air, a stable **climate**, and a steady supply of fresh water in the form of rain and snow.

THE ATMOSPHERE

HOW THE ATMOSPHERE WORKS

The atmosphere is a thin blanket of air that surrounds our planet. If Earth had no atmosphere, animals wouldn't be able to breathe and plants wouldn't be able to grow. Without a warming layer of air above it, Earth's surface would be freezing and all the world's water would turn to ice. There would be no wind, clouds, or rain – in fact, no weather at all, but just the Sun rising and setting every day.

OZONE LAYER

The air in the atmosphere consists mostly of two gases: nitrogen (78%) and oxygen (21%). Oxygen makes air breathable. It also shields us from the Sun's harmful ultraviolet radiation, which is absorbed by a form of oxygen called ozone. This gas is found in the stratosphere (second layer of the atmosphere).

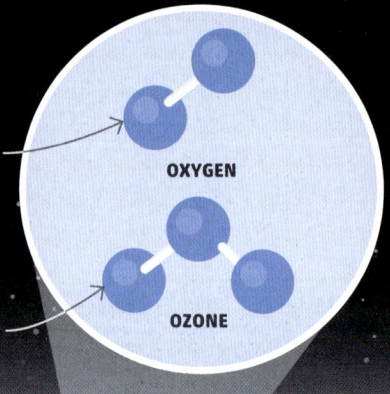

Normal oxygen molecules have two atoms each.

OXYGEN

Ozone molecules have three atoms each.

OZONE

Earth's horizon

▼ EARTH'S ATMOSPHERE FROM SPACE

In this picture, the atmosphere shows up as a blue haze that gradually gets darker with height. There is no clear edge to the atmosphere – the air just gets thinner until there is nothing but empty space.

Most clouds stay in the lowest layer of the atmosphere – the troposphere.

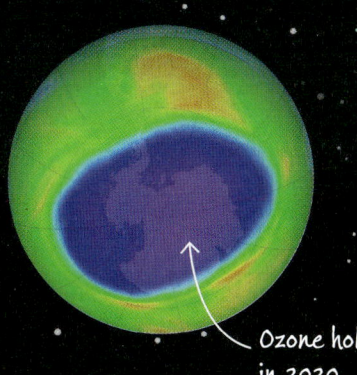

Ozone hole in 2020

THE OZONE HOLE

In the 1970s, scientists discovered that the ozone layer was very thin over Antarctica. This "ozone hole" was caused by chemicals called CFCs, used in fridges and aerosol cans. After CFCs were banned, the hole began to slowly heal.

STRATOSPHERIC CLOUDS

The stratosphere usually has no clouds. However, multicoloured clouds of glittering ice crystals occasionally form in cold polar regions. They are best seen just after sunset when the sky darkens but the high clouds catch the last rays of light.

LAYERS OF THE ATMOSPHERE

The atmosphere is divided into five layers, mainly defined by how the temperature changes within them.

EXOSPHERE
700–10,000 km (440–6,200 miles)

The outer edge of the atmosphere, where it fades into space, is called the exosphere.

THERMOSPHERE
80–700 km (50–440 miles)

In theory, this is the hottest layer of the atmosphere as the intense sunlight can warm air molecules to 1,500°C (2,700°F) in the daytime. However, there are so few air molecules here that a human visitor would find it freezing cold. The International Space Station and many satellites orbit Earth in this layer.

MESOSPHERE
50–80 km (31–50 miles)

The mesosphere is the coldest layer, with temperatures falling as low as –85°C (–120°F) at night. Meteors colliding with Earth burn up in the mesophere, creating shooting stars.

STRATOSPHERE
12–50 km (7–31 miles)

Planes and weather balloons fly in the stratosphere, which is usually cloud-free. The temperature rises with height in this layer because of ozone gas, which absorbs energy from the Sun.

TROPOSPHERE
0–12 km (0–7 miles)

This thin layer is where we live. It is the warmest and densest part of the atmosphere and is rich in water vapour from the oceans. Nearly all clouds and weather systems occur here.

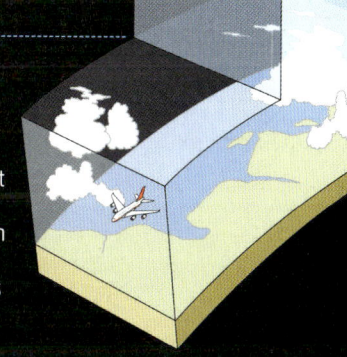

WHAT IS AIR MADE OF?

What looks like empty space around us is actually full of matter: air. The air in Earth's atmosphere is a mixture of colourless and odourless gases. We can't see or smell them, but we feel them every time the wind blows on our skin. Air is essential to life on Earth. It keeps the climate stable, and it provides the oxygen that living things need to release chemical energy from food.

▶ THE AMOUNT OF OXYGEN IN AIR

This experiment shows how much of the gas oxygen is in air. Wet iron wool is placed in an inverted glass tube with a dish of water at the bottom. When iron is exposed to oxygen, a chemical reaction takes place. The iron reacts with oxygen to form rust. This removes oxygen from the air.

NITROGEN 78%

OXYGEN 21%

ARGON 0.93%
CARBON DIOXIDE 0.038%
NEON 0.0018%
HELIUM 0.0005%
KRYPTON 0.0001%
HYDROGEN 0.00005%
XENON 0.000009%

WHAT GASES MAKE UP AIR?

Two gases make up 99 per cent of the air in Earth's atmosphere: nitrogen (78 per cent) and oxygen (21 per cent). The remaining 1 per cent includes tiny amounts of many other gases. Among them are carbon dioxide, which acts as a greenhouse gas, keeping Earth warm enough for life, and ozone, which protects life on Earth from harmful ultraviolet rays from the Sun.

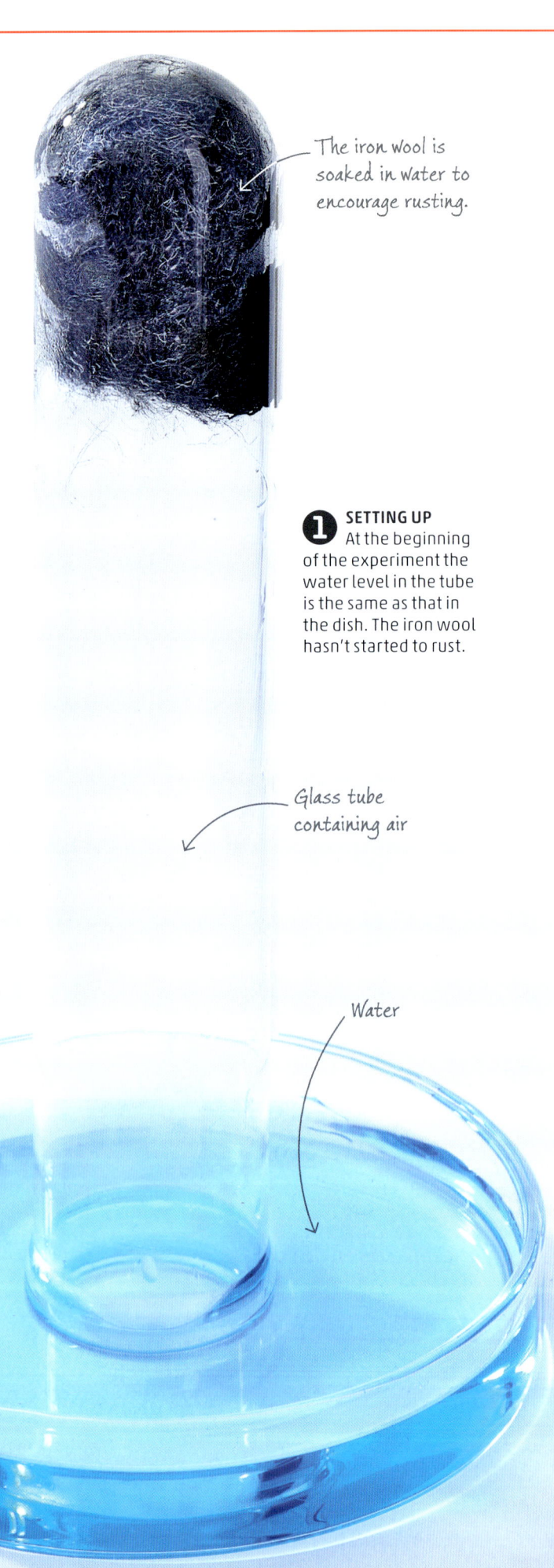

The iron wool is soaked in water to encourage rusting.

1 SETTING UP
At the beginning of the experiment the water level in the tube is the same as that in the dish. The iron wool hasn't started to rust.

Glass tube containing air

Water

The iron wool has rusted, using up all the oxygen gas in the tube.

2 TWO DAYS LATER
The water has risen because all the oxygen gas in the tube has turned into rust. The water level is about 21 per cent higher as air is 21 per cent oxygen.

RUSTY IRON
Reddish brown in colour, rust is scientifically known as iron oxide (Fe_2O_3). Water acts as a catalyst, speeding up the chemical reaction between iron and oxygen.

The water has risen up the tube by about 21 per cent.

WHAT ELSE IS IN AIR?
Gases aren't the only things in the atmosphere. Water in liquid or solid form and tiny particles like dust, soot, and ash float around. Living things can be found in the air too.

WATER
Clouds are made of liquid water droplets or ice crystals that drift in the air. Our eyes can't see the individual drops and crystals, but there are many billions of them in even a very small cloud.

AIRBORNE PARTICLES
Dust and other airborne particles have many sources. Volcanoes spew out vast amounts of ash; strong winds whip up grains of sand into the air; and factories release soot from their chimneys.

AERIAL PLANKTON
Tiny life forms float in the air. Known as aerial plankton, they include viruses, bacteria, pollen grains (above), and seeds. Some baby spiders travel hundreds of miles in the air, using a silk thread to catch the wind and find new homes.

HOW AIR PRESSURE WORKS

The air in Earth's atmosphere is not weightless. Pulled by gravity, it exerts a force on the planet's surface. This force is called air pressure. When we measure air pressure, we are weighing the air in the atmosphere all the way up to the edge of space. Every square centimetre of Earth's surface has about 1 kg of air pressing on it.

▼ **CRUSHING POWER**
The force that air pressure exerts around us can be shown in this simple demonstration. The air pressure inside a sealed bottle is lowered by cooling until the bottle's walls can no longer withstand the power of the air pushing on the outside.

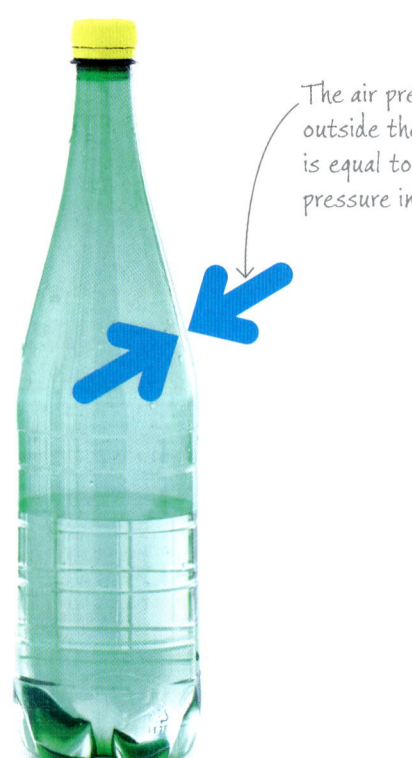

The air pressure outside the bottle is equal to the air pressure inside it.

Air pressure is now weaker inside the bottle than outside.

Water vapour condenses into droplets, reducing the amount of gas in the bottle.

1 ADDING WARM WATER
A plastic bottle is half-filled with hot water and left to stand for a minute so the air inside it warms up. The bottle is then sealed. The warm air inside the bottle is less dense than air outside but the pressure is equal.

2 COOLING THE BOTTLE
The bottle is plunged into a bowl of ice. This cools the air inside the bottle, which reduces the pressure. Water vapour condenses, reducing the pressure further.

The difference in air pressure is now so great that the sides of the bottle are crushed.

PRESSURE AND ALTITUDE
Air pressure falls as height above sea level increases. On mountains, the pressure may be less than half its value at sea level, which makes breathing difficult. That's why mountaineers sometimes carry oxygen tanks.

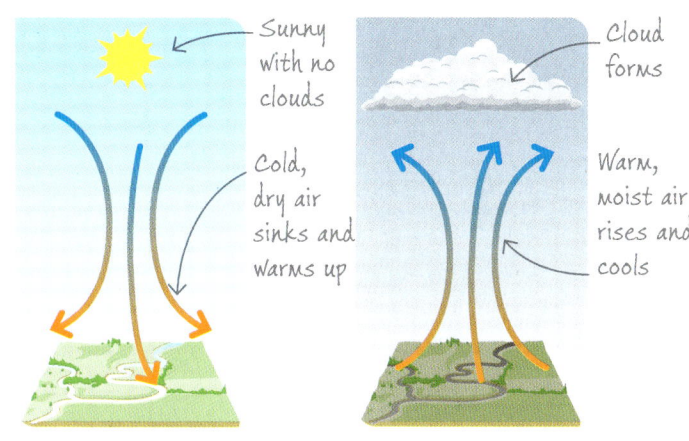

Sunny with no clouds

Cold, dry air sinks and warms up

Cloud forms

Warm, moist air rises and cools

HIGH PRESSURE   **LOW PRESSURE**

PRESSURE AND WEATHER
When air pressure is high, it usually means that cold, dry air is sinking from high in the sky. This prevents clouds forming and so brings fine weather. When air pressure is low, warm air near the ground rises and cools. Water vapour in the air condenses to make clouds and perhaps rain. Low air pressure usually means bad weather.

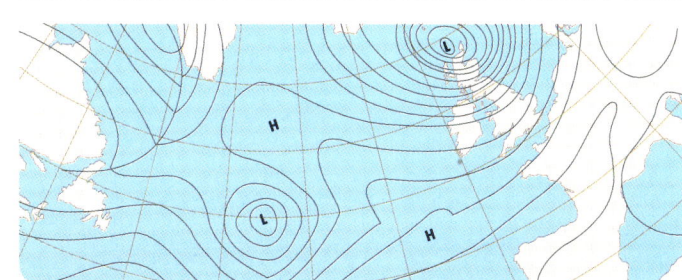

WEATHER MAPS
Weather maps used in forecasts often show the air pressure as this tells us what the weather will be like. Black lines called isobars join points of equal air pressure. Areas of very high pressure are indicated by the letter **H**, and very low areas by **L**.

3 CRUSHED BOTTLE
Air pressure continues to fall inside the bottle as it sits in the ice. Eventually, the bottle's sides cannot withstand the air pressure from outside and the bottle is crushed.

BEAUFORT SCALE

Long ago there were few scientific instruments to measure wind speed. So in 1805 the Beaufort Scale was devised to describe wind strength based on its effects on trees, buildings, and waves. The scale is still used today, with added wind speeds.

0 CALM
Smoke rises vertically.

**Wind 0 kph
(0 mph)**

1 LIGHT AIR
Smoke drifts with the wind.

**2–5 kph
(1–3 mph)**

2 LIGHT BREEZE
Wind felt on face.

**6–11 kph
(4–7 mph)**

3 GENTLE BREEZE
Leaves and small twigs move.

**12–19 kph
(8–12 mph)**

4 MODERATE BREEZE
Fallen leaves lifted from the ground.

**20–28 kph
(13–17 mph)**

HOW THE
WIND
BLOWS

Wind is air moving around in the atmosphere. The air flows from areas of high air pressure to areas of low pressure, similar to the way rivers flow from high ground to lower ground. The direction of winds is also strongly influenced by the rotation of Earth.

▶ WANDERING WINDS

In most parts of the world the wind direction is very variable. Over the tropical oceans, however, the winds most often blow from east to west. In the days of sailing ships, these reliable easterly winds were known as the trade winds because they helped sailors move cargoes around the world.

Sails catch the wind to propel the ship forwards.

5 **FRESH BREEZE**
Small trees begin to sway.

29–38 kph
(18–23 mph)

6 **STRONG BREEZE**
Large branches move.

39–49 kph
(24–30 mph)

7 **HIGH WIND**
Whole trees move.

50–61 kph
(31–38 mph)

8 **GALE**
Twigs break off trees.

62–74 kph
(39–46 mph)

9 **STRONG GALE**
Slight damage to roofs.

75–88 kph
(47–54 mph)

10 **STORM**
Trees uprooted; buildings damaged.

89–102 kph
(55–63 mph)

11 **VIOLENT STORM**
Lots of damage to buildings.

103–117 kph
(64–72 mph)

12 **HURRICANE FORCE**
Devastating damage.

118+ kph
(73+ mph)

HARNESSING THE WIND

For centuries, people harnessed the wind to grind wheat into flour in windmills. Today, wind turbines are used to generate electricity. They work best in windy places, such as hilltops or the sea.

JET STREAMS

About 10–15 km (6–9 miles) above Earth's surface are narrow bands of very strong winds called jet streams, which can exceed 240 kph (150 mph). Aircraft use these winds to speed up their journeys.

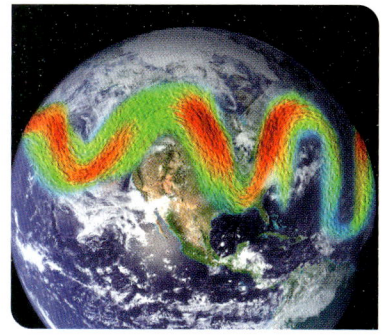

CORIOLIS EFFECT

Earth's rotation deflects winds flowing along the surface. This deflection is known as the Coriolis effect, and you can demonstrate it with a spinning globe.

1 Spin the globe anticlockwise to represent Earth's rotation. Then use a pen to quickly draw a line vertically down from north to south.

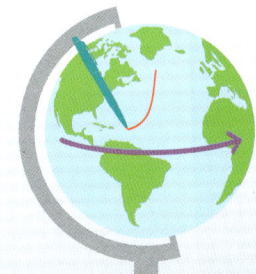

2 The line is not straight. It's deflected to the west as the globe rotates, creating a curve. The same thing happens to Earth's trade winds, which is why they blow from east to west.

PREVAILING WINDS

Some winds tend to blow in a particular direction across Earth's surface. These are called prevailing winds and influence the local climate. Westerlies, for example, come from the west and are warm and damp. Surface winds are linked to circulation cells that cycle air high in the atmosphere.

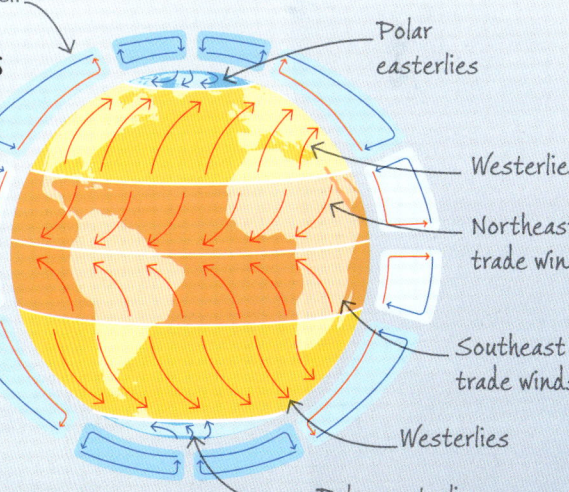

Circulation cell

Polar easterlies

Westerlies

Northeast trade winds

Southeast trade winds

Westerlies

Polar easterlies

HOW HOT AND COLD CLIMATES WORK

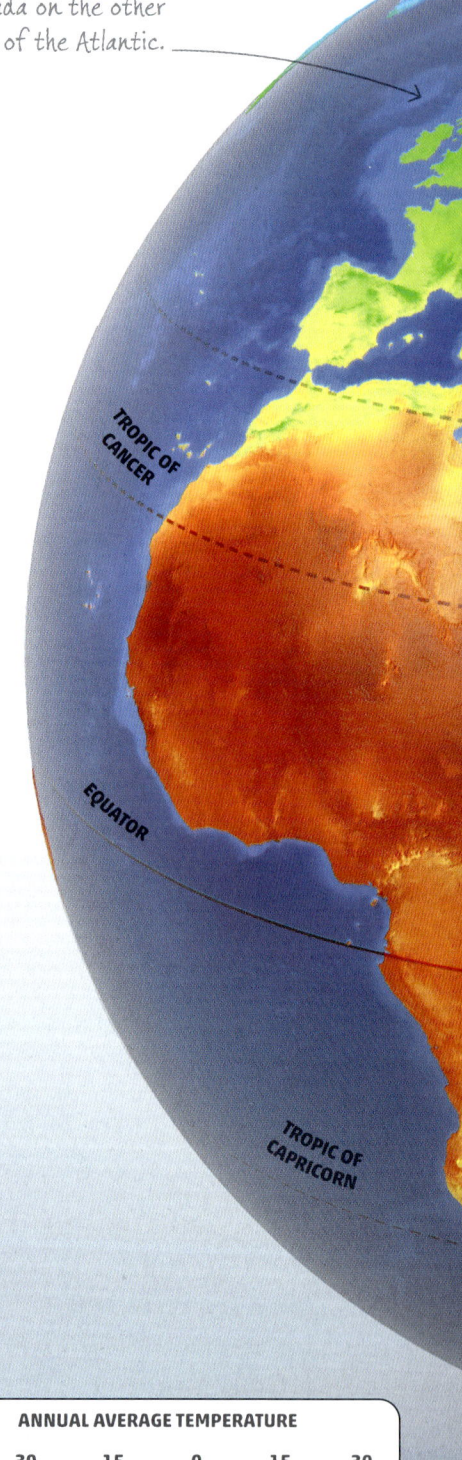

A warm ocean current keeps northwestern Europe warmer than Canada on the other side of the Atlantic.

TROPIC OF CANCER

EQUATOR

TROPIC OF CAPRICORN

Some places are scorching hot all year round, others are always freezing, and some places have warm summers yet chilly winters. Such long-term weather patterns are known as climates. We get different climates because of Earth's spherical shape and the way it receives heat from the Sun.

▶ AVERAGE TEMPERATURES
This globe shows the temperatures across Earth, averaged for the whole year. The warmest parts are in the tropical regions, near the equator. The coldest areas are at the poles.

UNEVEN WARMING
Because Earth's surface is curved, heat from the Sun is spread over a much larger area near the poles than at the equator. So climates are warmer near the equator, where the same amount of solar energy is focused on a smaller area.

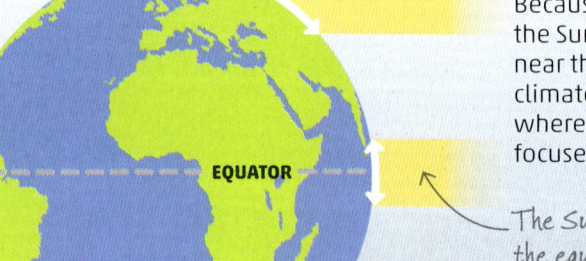

EQUATOR

The Sun's rays strike the equator straight on.

The Sun's rays strike polar regions at a shallow angle because Earth's surface curves away.

CLIMATE ZONES
Earth can be split broadly into four climate zones in each hemisphere. These different zones are separated by imaginary circles drawn round Earth. Tropical regions, for example, lie between the Tropics of Cancer and Capricorn.

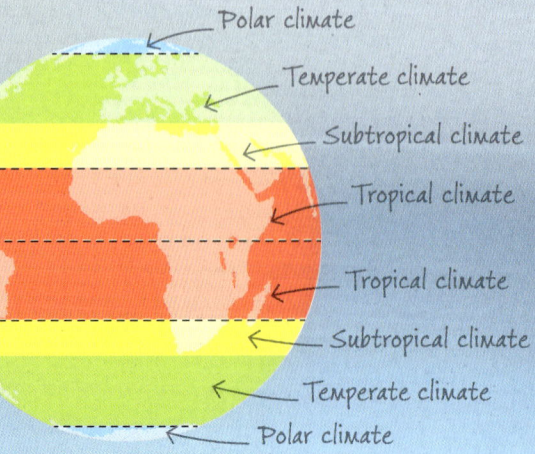

Polar climate

Temperate climate

Subtropical climate

Tropical climate

Tropical climate

Subtropical climate

Temperate climate

Polar climate

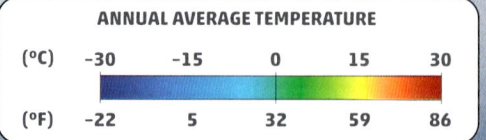

ANNUAL AVERAGE TEMPERATURE				
(°C) −30	−15	0	15	30
(°F) −22	5	32	59	86

ARCTIC
CIRCLE

POLAR CLIMATE
Little heat from the Sun reaches polar regions, even though the summer sun never sets. The climate is bitterly cold all year round, especially in winter when there is no sunlight at all.

◄ SVALBARD, NORWAY

This high plateau in Asia is about 4,000 m (13,000 ft) above sea level, so this area is very cold.

TEMPERATE CLIMATE
Places halfway between the equator and the poles have warm summers and cool winters. The climate in these regions is temperate (mild), being neither very hot nor very cold.

◄ BORJOMI-KHARAGAULI NATIONAL PARK, GEORGIA

SUBTROPICAL CLIMATE
Subtropical areas have long, hot summers and short, mild winters. Many of the world's major deserts lie in the subtropical zone.

◄ MEDINA PROVINCE, SAUDI ARABIA

TROPICAL CLIMATE
At the equator there is a lot of heat from the Sun warming the air and making it rise. There is plenty of moisture in the air, which condenses into clouds. So the climate in tropical regions is hot but also quite cloudy and wet.

◄ BWINDI IMPENETRABLE FOREST, UGANDA

HOW WET AND DRY CLIMATES WORK

The winds that blow around Earth pick up moisture from the oceans, and then dump it as rain or snow over land. Not all regions receive the same amount of rainfall, though. Deserts get very little rainfall, while some regions receive more rainfall in a single day than deserts receive in a year.

▶ WET AND DRY

This globe shows average annual rainfall in different parts of the world. The wettest places (dark blue) are near the Equator. Deserts, on the other hand, which are the driest places (light blue), are mostly just north or just south of the wet zones, along the Tropics of Cancer and Capricorn.

ANNUAL PRECIPITATION				
MM 0	2,500	5,000	7,500	10,000
IN 0	98	197	295	394

WHY IS IT WET NEAR THE EQUATOR?

Air cycles known as Hadley cells cause wet regions at the Equator and dry regions around it.

1 Moisture-laden trade winds converge at the Equator. The warm climate heats the humid air and makes it rise.

2 As the air rises, it cools and water vapour condenses to create clouds and rain, which falls back to Earth.

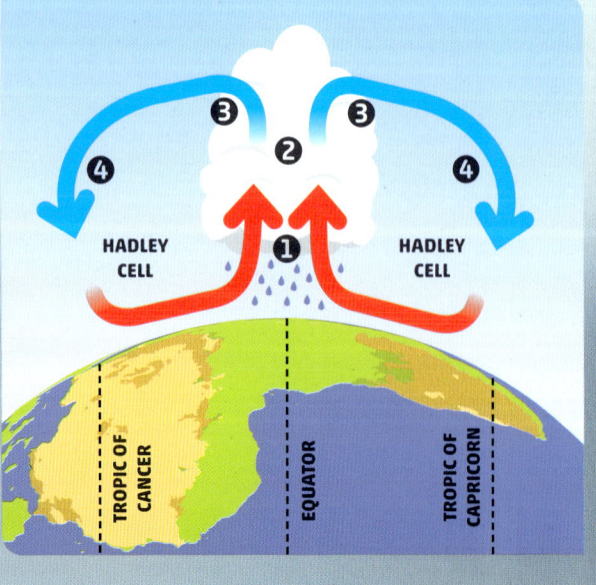

HADLEY CELL

HADLEY CELL

TROPIC OF CANCER

EQUATOR

TROPIC OF CAPRICORN

3 Having cooled and lost its moisture, the air is now dry. It travels away from the Equator, spreading north and south.

4 The cool, dry air sinks back over the Tropics of Cancer and Capricorn, giving these areas a dry climate. The air then flows back to the Equator and the cycle repeats itself.

DESERTS
Deserts have an annual rainfall of less than 25 cm (10 in), and they can go for months or even years without any rainfall. Animals and plants need special adaptations to survive in deserts.

RAINFORESTS
Around the Equator, it rains nearly every day, and annual rainfall can exceed 200 cm (79 in). The warm climate and plentiful rain are perfect for plants, so rainforests flourish around the Equator.

WET AND DRY SEASONS
Tropical areas between the desert zone and the Equator tend to have distinct wet and dry seasons. Plants die or become dormant in the dry season, and the land turns dry and dusty. When rain returns, the landscape turns green again.

TROPIC OF CANCER

EQUATOR

TROPIC OF CAPRICORN

DRIEST DESERT
Not all of the world's deserts are in hot places. Antarctica is considered a "polar desert" because it gets only about 50 mm (2 in) of precipitation a year. This falls as snow and piles up without melting because it's so cold. The South Pole is 2,830 m (9,285 ft) above sea level, but this is mostly just a thick sheet of ice.

RAIN BELT
A permanent belt of cloud and rain wraps around the Equator. Humid trade winds that blow across the oceans, flowing towards the Equator, constantly add moisture to this belt. Because Earth's axis is tilted, the rain belt shifts north in the northern summer, and south in the northern winter, creating a cycle of wet and dry seasons that repeats every year.

HOW OCEAN CURRENTS WORK

The oceans play an enormous role in Earth's climate, absorbing heat from the Sun and moving that heat around in flows of water called currents. These currents are much slower than winds, typically travelling less than half a metre (1.5 ft) per second at the surface, and even less than that further down.

Warmer water appears red or yellow on this satellite image, while cold water appears blue.

▼ THE GULF STREAM

This image, created from satellite data, shows a current of warm water called the Gulf Stream flowing out of the Gulf of Mexico and towards Europe, carrying heat northwards. The current is driven by wind and by the sinking of cold, salty ocean water near the North Pole. Because of the Gulf Stream, northwestern Europe has a relatively mild climate for its latitude.

Warm water flows out of the Gulf of Mexico.

GLOBAL CURRENTS

The map shows the ocean currents around the globe. The largest oceans have circulating currents called gyres, caused by surface winds. The gyres flow clockwise in the northern hemisphere and anticlockwise in the southern hemisphere. A cold current also flows all the way around the southern hemisphere, uninterrupted by land. It's powered by strong westerly winds called the Roaring Forties, which blow between 40° and 50° latitude.

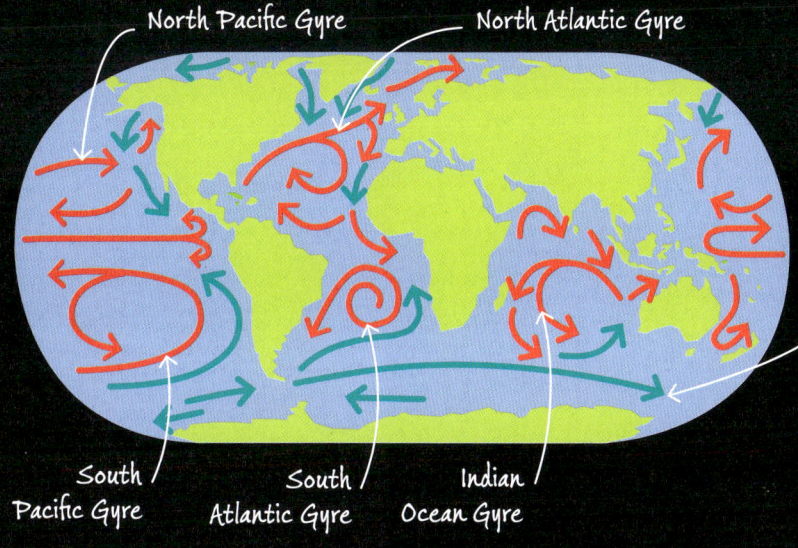

North Pacific Gyre

North Atlantic Gyre

The Antarctic Circumpolar Current flows around the globe, driven by the Roaring Forties.

South Pacific Gyre

South Atlantic Gyre

Indian Ocean Gyre

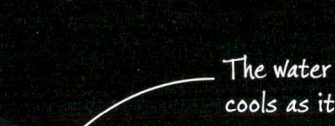

■ WARM CURRENTS
■ COLD CURRENTS

The water cools as it moves north.

These swirling currents are called eddies.

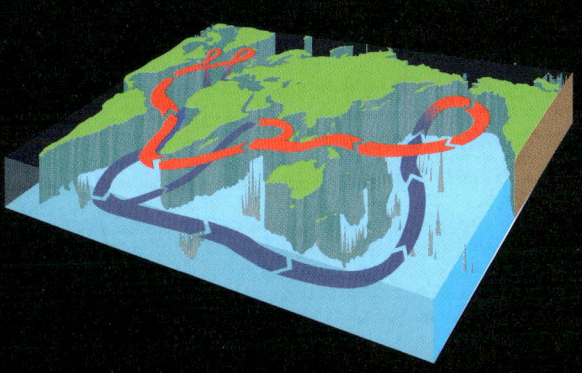

GLOBAL CONVEYOR BELT

Surface currents combine with deep underwater currents to make a "global conveyor belt" of water, which connects all the world's oceans. The conveyor belt is slower than wind-driven currents, taking about 1,000 years to complete a full cycle. The underwater currents start near the North Pole, where cold, salty water sinks down and begins its slow and winding journey around the world.

FLOATING GARBAGE

The "Great Pacific Garbage Patch" is a vast area of floating trash that has been slowly trapped inside the swirling currents of the North Pacific Gyre. It contains large discarded items, like fishing nets and plastic packaging, as well as trillions of tiny plastic particles that are too small to see (microplastics).

DEBRIS WASHED UP ON BEACH, HAWAII

Thin, high clouds form first as a cold front approaches.

FRONTS FROM SPACE
Using satellite imagery, we can view weather fronts from space. They show up as long bands of clouds, some of which can be over 1,000 km (625 miles) long. Satellite images help us to study weather patterns, and predict where storms are about to hit based on the movement of the clouds.

HOW
WEATHER
FRONTS
WORK

The air in Earth's atmosphere is not evenly mixed.
Sometimes, a huge volume of air collides with another volume of air that's warmer, colder, wetter, or drier. The boundaries between these colliding air masses are called weather fronts, and they are like battlegrounds. They often bring bad weather, including clouds, rain, snow, and thunderstorms.

▲ COLD FRONT
An approaching cold front brings dramatic changes. Thin, high clouds form first, as warm air is pushed upwards. The clouds soon grow thicker and darker, as they block out more light from the Sun. This is a sign that rain is on the way.

WEATHER MAPS

Weather maps use coloured symbols to show weather fronts, and lines called isobars to show areas of equal pressure. Areas of high pressure bring settled and dry weather. Areas of low pressure, on the other hand, bring unsettled weather that's often wet and windy.

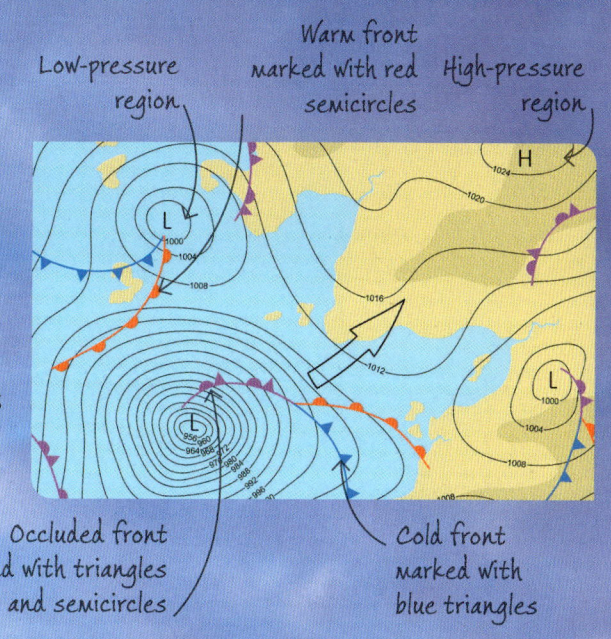

Low-pressure region

Warm front marked with red semicircles

High-pressure region

Occluded front marked with triangles and semicircles

Cold front marked with blue triangles

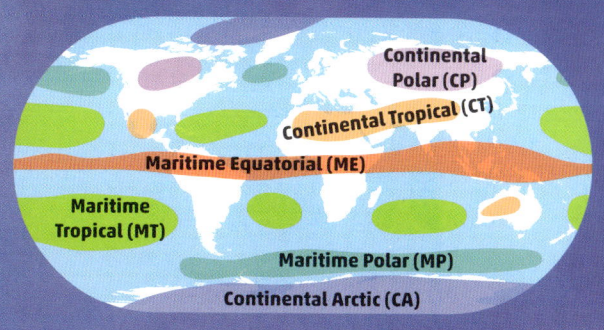

PRINCIPAL WORLD AIR MASSES

Continental Polar (CP)

Continental Tropical (CT)

Maritime Equatorial (ME)

Maritime Tropical (MT)

Maritime Polar (MP)

Continental Arctic (CA)

AIR MASSES

Whether air masses are warm, cold, dry, or wet depends mainly on where they come from. Polar air masses are cold, while tropical air masses are warm. Air over the ocean (maritime air masses) is humid, while continental air is dry. The map above shows the usual pattern of major air masses.

TYPES OF WEATHER FRONT

There are three main types of weather front, and each one causes a particular kind of weather.

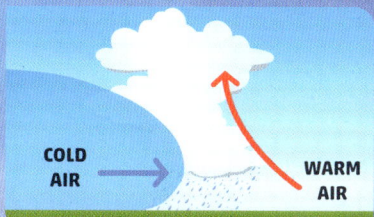

COLD AIR

WARM AIR

Cold front
As cold, dense air moves forwards, it pushes the lighter, warmer air upwards, making a band of thick cloud with often heavy rain.

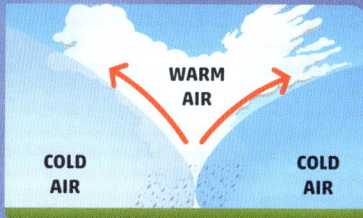

WARM AIR

COLD AIR

COLD AIR

Occluded front
At an occluded front, a mass of warm air is squeezed upwards by two colder air masses coming in from either side. Clouds form in the rising warm air.

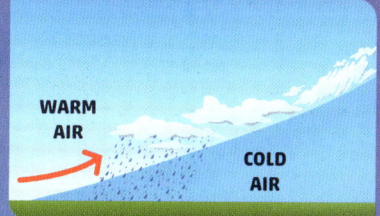

WARM AIR

COLD AIR

Warm front
Warm air moving forwards will rise above a colder, denser air mass, forming a wide band of cloud with rain and drizzle.

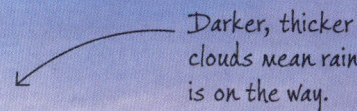

Darker, thicker clouds mean rain is on the way.

HOW HURRICANES WORK

Hurricanes are swirling masses of deep clouds, heavy rain, and extremely strong winds. They produce some of the most destructive weather on Earth. Hurricanes occur in the tropical regions and always form over the ocean. In the Atlantic and east Pacific they are called hurricanes but in the west Pacific they are known as typhoons, and over the Indian Ocean they are known as cyclones.

Air spins around the eye in the centre.

High in the atmosphere, wind spirals around the storm.

Bands of cloud and rain form where warm air rises.

Air is drawn in, producing strong winds.

INSIDE A HURRICANE
Hurricanes are fuelled by heat from the ocean that causes rapidly rising air and very strong winds. Air is sucked towards the centre of the storm just above the ocean surface, rises in bands, and spirals out at the top.

Spiral cloud bands

Eye wall

Eye

EYE OF A HURRICANE
The eye of a hurricane is relatively calm, but the ring of cloud around it – the eye wall – is where the strongest winds are. If the eye of a hurricane passes over you, you'll experience the raging winds of the eye wall twice, with a period of fair weather between.

HURRICANE DAMAGE

Hurricanes can cause catastrophic damage in three main ways: through violent winds, through storm surges from the sea, and from torrential rain.

STRONG WINDS
A category 5 storm can produce gusts of over 252 kph (157 mph). The winds in a category 5 storm can flatten buildings, trees, and power lines.

STORM SURGE
Low air pressure at the centre of a hurricane causes the sea level to rise beneath it, leading to huge waves and coastal flooding.

HEAVY RAIN
Hurricanes produce huge amounts of rain. The winds weaken if the storm moves inland, but the heavy rain can still cause severe flooding.

▼ HURRICANE FLORENCE

This photograph, taken from the International Space Station in 2018, shows Hurricane Florence raging over the Atlantic Ocean. The Atlantic hurricane season starts at the beginning of June and finishes at the end of November, with an average of six hurricanes developing each year. In the Atlantic, the names of the hurricanes go from A to Z through the year. They alternate between male and female names, with Q, U, X, Y, and Z left out as there are few names starting with those letters.

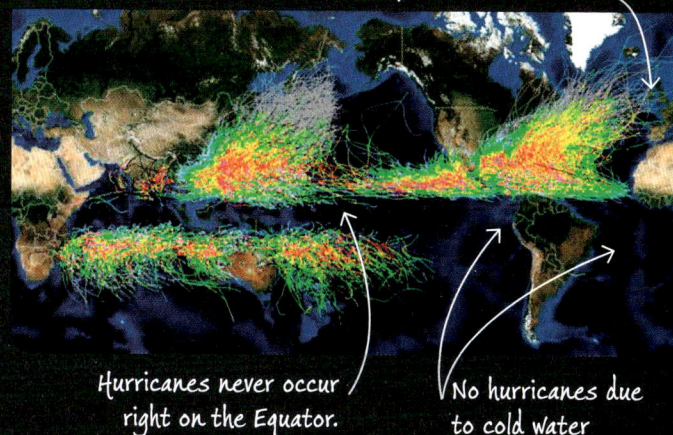

Storms weaken as they track northwards.

Hurricanes never occur right on the Equator.

No hurricanes due to cold water

WHERE DO HURRICANES HAPPEN?

This map shows the track of every hurricane from 1848 to 2013. The region with the most storms is the west Pacific. There are no hurricanes in the southeast Pacific and only one in the South Atlantic because the water there is usually too cold. The track of each hurricane is shown as a line, with the colour showing the strength of the storm.

HURRICANE SCALE

The strength of a hurricane is rated on the Saffir-Simpson scale, which is based on wind speed. Category 5 hurricanes are rare, with only four making landfall in the US between 1935 and 2018.

1 119–153 kph (74–95 mph)
Weak trees fall, roof tiles blown off

2 154–177 kph (96–110 mph)
Houses damaged, trees snapped

3 178–208 kph (111–129 mph)
Trees uprooted, major damage to homes

4 209–251 kph (130–156 mph)
Roofs blown off; no water and electricity

5 >252 kph (>157 mph)
Catastrophic damage; area uninhabitable

HOW TORNADOES WORK

A tornado is a whirling funnel of cloud and wind that can hurl a car into the air and destroy a building in seconds. The average tornado touches the ground for a matter of minutes, but it can leave a trail of devastation in its wake.

HOW TORNADOES FORM

Tornadoes are rare as they form only under certain specific conditions. They develop from powerful storm clouds called supercells.

1 CLOUDS FORM
For a tornado to form, the air near the ground must be hot and humid, with cooler air above. The hot air rises, forming towering thunderclouds several kilometres tall.

WARM AIR

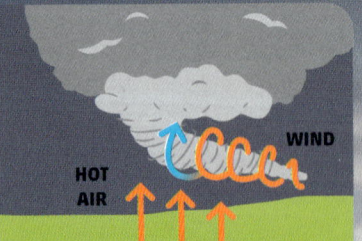

2 ROTATION BEGINS
Air drawn into the cloud begins to rotate. At first the rotating air moves horizontally.

HOT AIR   WIND

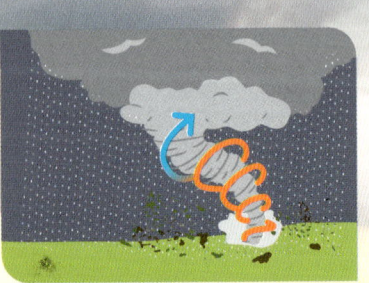

3 TOUCHDOWN
Strong winds and heavy rain concentrate the rotating air into a tight funnel and turn it upright. When this funnel hits the ground a tornado has formed.

▼ FURIOUS FUNNEL
Just as water spirals around as it drains through a plughole, the air rushing up into a tornado forms a violent, spinning vortex. A newly formed tornado consists of tiny water droplets and is usually white or grey, but the base changes colour as it picks up earth and other debris.

Base of thundercloud

TORNADO SCALE

Tornadoes are classified from 0 to 5 using the Enhanced Fujita (EF) Scale, which is based on how much damage they do.

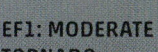
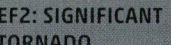

EF0: GALE TORNADO
Some damage to chimneys. Branches snapped off trees.

EF1: MODERATE TORNADO
Tiles blown off roofs and cars pushed off the road.

EF2: SIGNIFICANT TORNADO
Roofs detached from houses and trees snapped or uprooted.

EF3: SEVERE TORNADO
Considerable damage to buildings. Cars lifted off the ground.

EF4: DEVASTATING TORNADO
Trucks lifted off the ground. Houses completely flattened.

EF5: INCREDIBLE TORNADO
Significant damage to large buildings. Cars lifted high in the air.

Tornado funnel

Debris lifted by strong winds

KANSAS
OKLAHOMA
TEXAS

TORNADO ALLEY, USA

Tornadoes are rare events, but some places experience far more tornadoes than others. The USA has more tornadoes than any other country. Tornadoes are most common in the central states, with Texas receiving the most – an average of 136 a year.

THE TORNADO SWATH

The track of damage caused by a tornado is called the swath. The pale line across the image above shows a swath that is around 2.5 km (1.5 miles) wide at its widest point.

HOW CLOUDS FORM

Clouds may bring bad weather, but without them there would be no fresh water on Earth and life would not exist. As well as watering the land, clouds help regulate the climate. They cool Earth by reflecting sunlight back into space but can keep us warm at night by acting like a blanket to trap heat near the ground.

▶ THE WATER CYCLE

Clouds perform a vital job in Earth's water cycle – the movement of water on, below, and above the planet's surface. Water that evaporates from the ocean surface rises and cools until it forms clouds. These clouds are blown across land, where they produce rain or snow that feeds lakes and rivers and waters vegetation.

HOW RAIN CLOUDS FORM

Rain clouds form when warm, moist air rises. The rising air cools, making water vapour condense into droplets. There are three main ways that large masses of air can be lifted, creating rain clouds.

PRECIPITATION

Clouds release precipitation (rain, snow, or hail).

Some water seeps into the ground and slowly flows back towards the sea.

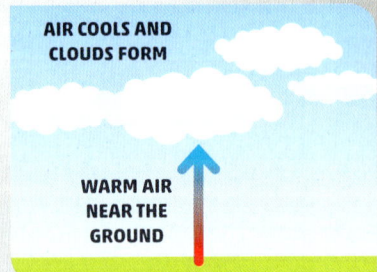

AIR COOLS AND CLOUDS FORM

WARM AIR NEAR THE GROUND

WIND MOVES UP AND OVER A HILL

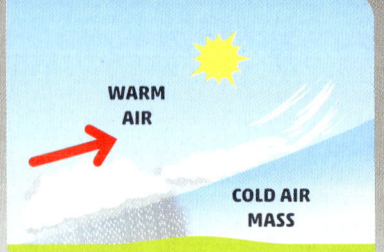

WARM AIR

COLD AIR MASS

① CONVECTIVE CLOUDS
When Earth's surface is warm, it heats the air above it, making it rise. This process is called convection. Convective clouds are common in summer and in the tropics. They produce brief showers or thunderstorms.

② OROGRAPHIC CLOUDS
When wind blows over a hill or a mountain, the air rises and cools, causing clouds to form. Most of the rain from the clouds falls on the windward side of the mountain, with the far side in a dry "rain shadow".

③ FRONTAL CLOUDS
When two large air masses collide, the warmer air mass rises over the colder, denser air and forms a large layer of cloud. The boundaries between large air masses are called fronts, and the rain produced by frontal clouds can be persistent and drizzly.

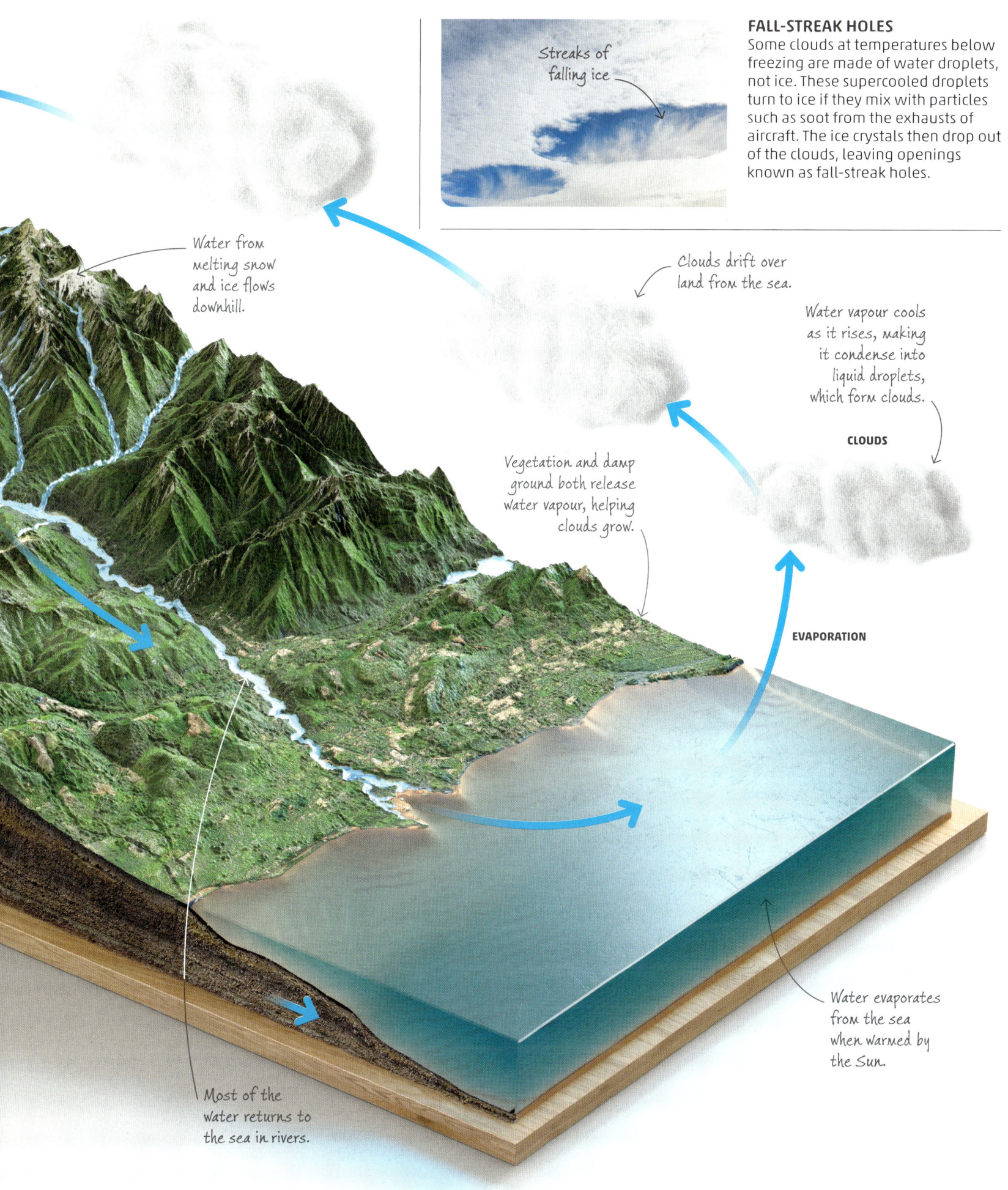

FALL-STREAK HOLES
Some clouds at temperatures below freezing are made of water droplets, not ice. These supercooled droplets turn to ice if they mix with particles such as soot from the exhausts of aircraft. The ice crystals then drop out of the clouds, leaving openings known as fall-streak holes.

Streaks of falling ice

Water from melting snow and ice flows downhill.

Clouds drift over land from the sea.

Water vapour cools as it rises, making it condense into liquid droplets, which form clouds.

CLOUDS

Vegetation and damp ground both release water vapour, helping clouds grow.

EVAPORATION

Water evaporates from the sea when warmed by the Sun.

Most of the water returns to the sea in rivers.

TYPES OF CLOUD

White and fluffy, thin and wispy, or dark and heavy, clouds are a good indicator of what is happening in the atmosphere. Their shape depends on how high they are and how much water they contain. The ten main cloud formations are divided into three groups according to their level in the atmosphere: low clouds, medium clouds, and high clouds.

CIRRUS
Cirrus are high, wispy clouds made of ice crystals. They appear in fair weather but also tend to form ahead of a warm front, which means rainy weather may be on the way.

CUMULONIMBUS
These towering clouds can reach 20 km (12 miles) tall and often spread out flat at the top. They produce torrential rain, hail, and thunder. One cumulonimbus cloud can store as much energy as 10 atom bombs.

Updraughts of warm, moist air carry the cloud high into the atmosphere.

ALTOCUMULUS
Altocumulus are mid-level clouds that form a variety of shapes, including small towers and flying saucers. They contain more water droplets than ice crystals, which makes them grey.

CIRROCUMULUS

These bouncy little clouds that spread across the sky sometimes resemble the scales of a fish, earning the nickname "mackerel sky". They contain ice crystals and supercooled water, and are usually associated with fair weather.

CIRROSTRATUS

Covering much of the sky, cirrostratus clouds hang high in the air like a transparent veil. They sometimes create circular haloes around the Sun, occasionally with bright spots called sundogs on either side. Their presence usually indicates the arrival of rain or drizzle the next day.

Sun

Sundog

ALTOSTRATUS

When the sky looks pale grey and flat, it is probably covered by altostratus clouds. The Sun can be seen weakly through these clouds, but it does not cast shadows. They signify that bad weather is on the way.

NIMBOSTRATUS

These occur when altostratus clouds deepen and thicken as they meet a warm air mass. They darken because of the large water droplets they contain, which eventually fall as rain, often for several hours.

STRATOCUMULUS

Stratocumulus clouds are the most common type of cloud. They are clumpy grey or white shapes with flat bottoms and spaces between them. Although usually associated with rain, they rarely produce more than drizzle.

STRATUS

These flat clouds cover the sky like a blanket and may be grey and featureless or ragged and broken. They are the lowest clouds and sometimes touch the ground to create fog or mist.

CUMULUS

Cumulus clouds are the cauliflower-shaped clouds with fluffy white tops that you often see on sunny days. They form when warm air rises from the ground and cools, the water vapour condensing into droplets. Cumulus clouds occasionally produce light showers.

SUPERCELL

Most tornadoes form in supercells – the most powerful kinds of thunderstorms. Supercell clouds can develop a rounded shape at the base as powerful updraughts spiral around inside them. These updraughts can reach speeds of 140 kph (90 mph), which is fast enough to suspend hailstones the size of grapefruits. As the rising air cools, its moisture condenses and releases heat energy, adding to the storm's power.

HOW RAIN WORKS

Rain can be a nuisance, but it's vital to life on Earth, providing fresh water to plants and animals. Rain forms when invisible water vapour in air rises and cools enough to condense into tiny droplets. If these droplets grow larger than around 1 mm (0.04 in) wide, they begin to fall.

RAINY SEASON

In many parts of the world, almost all the rain falls in one season. Rain pours down on Africa's savanna grassland during the wet season. But in the dry weather that follows, plants turn brown and shrivel, rivers and lakes dry up, and the parched earth cracks. Vast herds of zebra and wildebeest migrate in search of water and food. However, when the rain begins again, plants and trees revive, seeds germinate, and animals return.

WET SEASON

DRY SEASON

1 NEW RAINDROP
Tiny, newly formed raindrops are shaped into a sphere by a force called surface tension, which acts in the droplet's skin.

2 GROWING WIDER
When droplets become more than about 1 mm (0.04 in) wide, they start to fall. The force of the air pushing under them makes them lose their spherical shape.

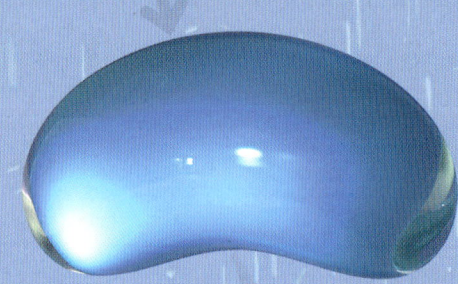

3 BUN SHAPE
The raindrop speeds up, falling at around 20 kph (12 mph). It forms a bun shape as air pushes into it. A cavity forms in the bottom but the top stays rounded thanks to surface tension.

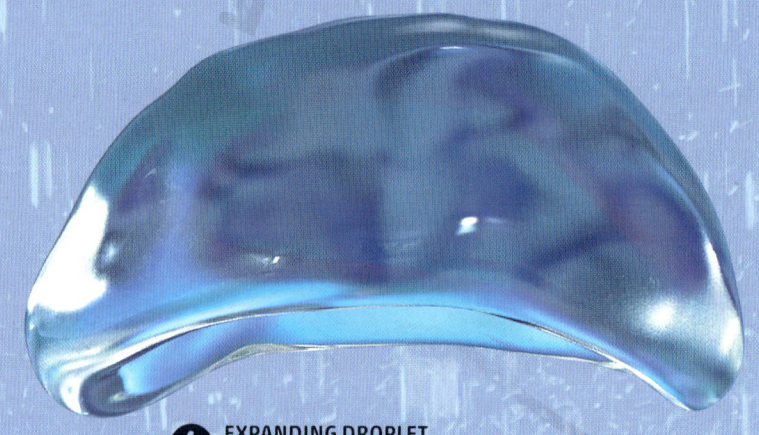

4 EXPANDING DROPLET
The force of the air pushing under the droplet continues to stretch it, increasing the size of the cavity too.

MEASURING RAIN

Rain is measured in millimetres or inches using a rain gauge. The simplest gauges are a measuring tube with a funnel or a wider opening at the top. The gauges show you how much rain falls on an area of land over a certain period of time.

CLOUDBURSTS

Occasionally, storm clouds dump a week's or a month's worth of rain in a matter of minutes. This is called a cloudburst and can cause flash floods. Cloudbursts often occur when thunderclouds pass over mountains. They are especially common in the Himalayas during the monsoon season.

DRY STORMS

Sometimes rain falls from a cloud but it doesn't reach the ground. It hits warm or dry air and evaporates. These dry storms can be seen in the distance as streaks descending from the base of a cloud.

The raindrop becomes bell-shaped, prior to breaking up.

◄ THE LIFE OF A RAINDROP

A raindrop begins to form when water vapour in the air condenses around a dust particle. Small and spherical at first, the droplets collide and grow into larger drops. When these become heavy enough to start falling, they begin to change shape, but they never form a teardrop shape. If a raindrop grows larger than 4–5 mm (0.2 in) wide, it splits into spherical drops again.

5 BELL SHAPE
Eventually, the air pushing into the raindrop from below creates an unstable bell shape.

6 SPLITTING
The bell breaks up into smaller droplets, which become spherical again due to surface tension. These either continue to fall as rain or are swept back up into the cloud.

▶ SPLITTING COLOURS

Sunlight is white to our eyes, but in reality it's a mixture of all the colours of the rainbow: red, orange, yellow, green, blue, indigo, and violet. When sunlight passes in and out of raindrops, these colours bend by different amounts and separate to form a rainbow.

1 LIGHT ENTERS RAINDROP
When sunlight enters a raindrop, it bends (refracts). The different colours that make up white light bend by varying amounts. Blue bends the most and red the least, with other colours in between.

SUNLIGHT

HOW RAINBOWS WORK

They appear out of nowhere and vanish just as quickly. Rainbows are the most colourful of all weather phenomena and happen when rain is falling but the Sun is shining. But you can only see one if you are standing in the right place at the right time.

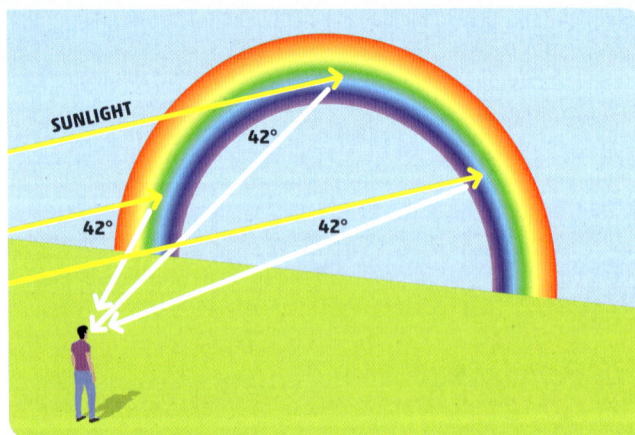

SUNLIGHT

42°

42°   42°

WHY ARCHES FORM

To see a rainbow, you must be standing with the Sun behind you and rain in front of you. The Sun also has to be fairly low in the sky. This is because bright colours are only visible when the angle between light rays hitting raindrops and rays bouncing back to you is around 42°. Raindrops in just the right place to reflect light at this angle form a circular arc. The bottom of the arc is blocked by the ground, so we normally see an arch shape.

3 LIGHT LEAVES RAINDROP
Light is refracted a second time as it leaves the raindrop. This makes the colours spread out even further.

The colours spread out and separate.

REFRACTION

ANGLE OF REFLECTION

Raindrops scatter and reflect light in all directions, but the strongest colours are seen where sunlight bounces back at an angle of 40–42°. Blue is seen at 40° and red at 42°.

42°

② REFLECTION

When light hits the back of a raindrop, some of it is reflected back towards our eyes. All the different colours are reflected.

REFLECTION

Raindrops act like tiny lenses, splitting light into colours.

SEEING DOUBLE

Look carefully and you might spot a weaker second rainbow above the main one. This happens because some of the light is reflected twice inside raindrops and bounces back to observers at an angle of 50–53°. The second reflection reverses the order of the colours.

FULL CIRCLE

Most rainbows are cut off by the horizon, but if you see one from very high up – such as from inside a plane – you might be lucky enough to see a full circle.

HOW MIST, FOG, AND DEW WORK

If you think clouds are something only birds and planes encounter at close range, you may be surprised to learn that you can walk through them. Mist and fog are simply clouds that have formed at ground level.

Because it contains lots of water droplets, fog is difficult to see through and feels damp.

Dubai's frequent winter fogs are caused by moist sea air flowing inland, where the ground is cold.

Heat radiates from surface

Moisture condenses into fog

▲ SHROUDED IN FOG

The city of Dubai, in the United Arab Emirates, lies swathed in fog on a winter morning. Like their sky-high cousins, ground-level clouds are made of tiny droplets of liquid water suspended in air. It is called fog if you can see less than 1 km (0.62 miles) in front of you. Mist is thinner than fog, so you can see further. Mist and fog clear as the day warms up and the droplets turn back into vapour.

HOW FOG FORMS

Fog can form in lots of different ways, but all involve air cooling down and water vapour condensing into droplets.

Radiation fog
On cold, clear nights, the land radiates (emits) heat it absorbed during the day. As the ground loses heat and cools down, it cools the air above. Water vapour in the air cools and condenses, causing radiation fog.

DEW
As objects cool at night, they may cool the air around them enough for airborne water vapour to condense as tiny beads of liquid – dew. A similar thing happens when you take a cold drink from the fridge on a hot day and water forms on the can.

LIFE-SUSTAINING FOG
Redwood trees growing in California, USA, rely on fog rolling in from the sea for their survival. The trees – the world's tallest, often exceeding 90 m (300 ft) in height – need huge amounts of water. In the dry summers, their leaves trap droplets of fog and absorb the water.

GHOSTLY ILLUSION
A Brocken spectre is a ghostly figure (a spectre) that appears in fog or mist. The "ghost" is really just the shadow of the person observing it while standing on a hill and looking into the mist, with the Sun behind them. A rainbow halo may also form around the spectre.

Warm, moist air cools as it moves over a cold surface — Fog forms

Cold, dense air sinks into valley

Cold air moves over warm, moist air

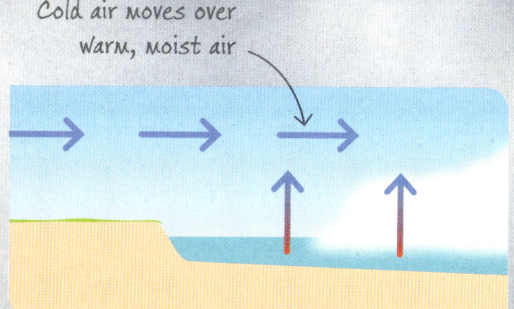

Advection fog
As warm, moist air moves across cold land or water, the air cools and fog forms. This type of fog is called advection fog and is common at sea and along coasts. It also occurs when a warm front passes over snow-covered land.

Valley fog
This type of fog occurs when cold air sinks to the bottom of a valley and gets trapped. Unlike other forms of fog, which often disappear quickly, valley fog may linger for days before clearing.

Evaporation fog
Warm water evaporating from a lake, pond, or moist land warms the air above and causes it to rise. As the warm, moist air mixes with cold air passing overhead, it cools and fog forms.

HOW HAILSTONES FORM

Hailstones are icy pellets that form in towering thunderclouds. Usually the water in thunderclouds falls as rain, but under certain atmospheric conditions it turns to ice and becomes hailstones.

The final coating of ice may be bumpy or spiky as smaller wet hailstones collide with and stick to the large hailstone.

INSIDE A STORM CLOUD

Hailstones form when water in a thundercloud becomes supercooled. This means the water gets down to below freezing point, but instead of turning to ice it stays liquid. Droplets of the supercooled water are swept to the top of the cloud by powerful updraughts and freeze. As they fall back down through the cloud they are coated with more supercooled water, which freezes instantly. After several rise-and-fall cycles, the ice pellets become too heavy to stay airborne and fall.

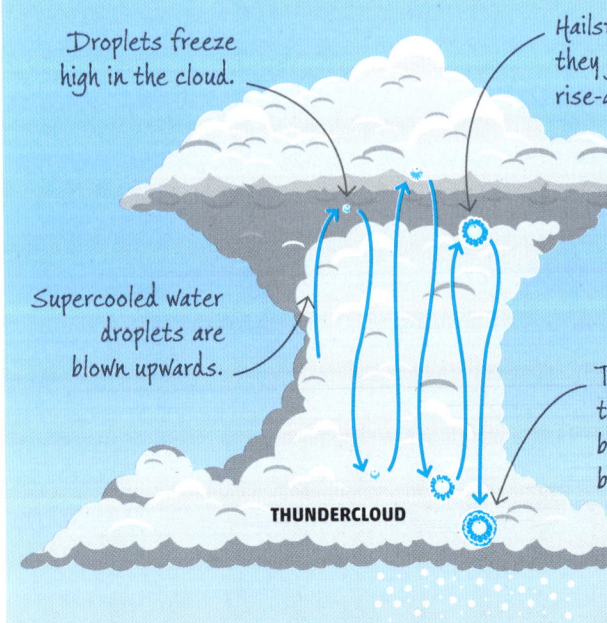

Droplets freeze high in the cloud.

Hailstones grow as they go through rise-and-fall cycles.

Supercooled water droplets are blown upwards.

The hailstone falls to the ground when it becomes too heavy to be blown back up.

THUNDERCLOUD

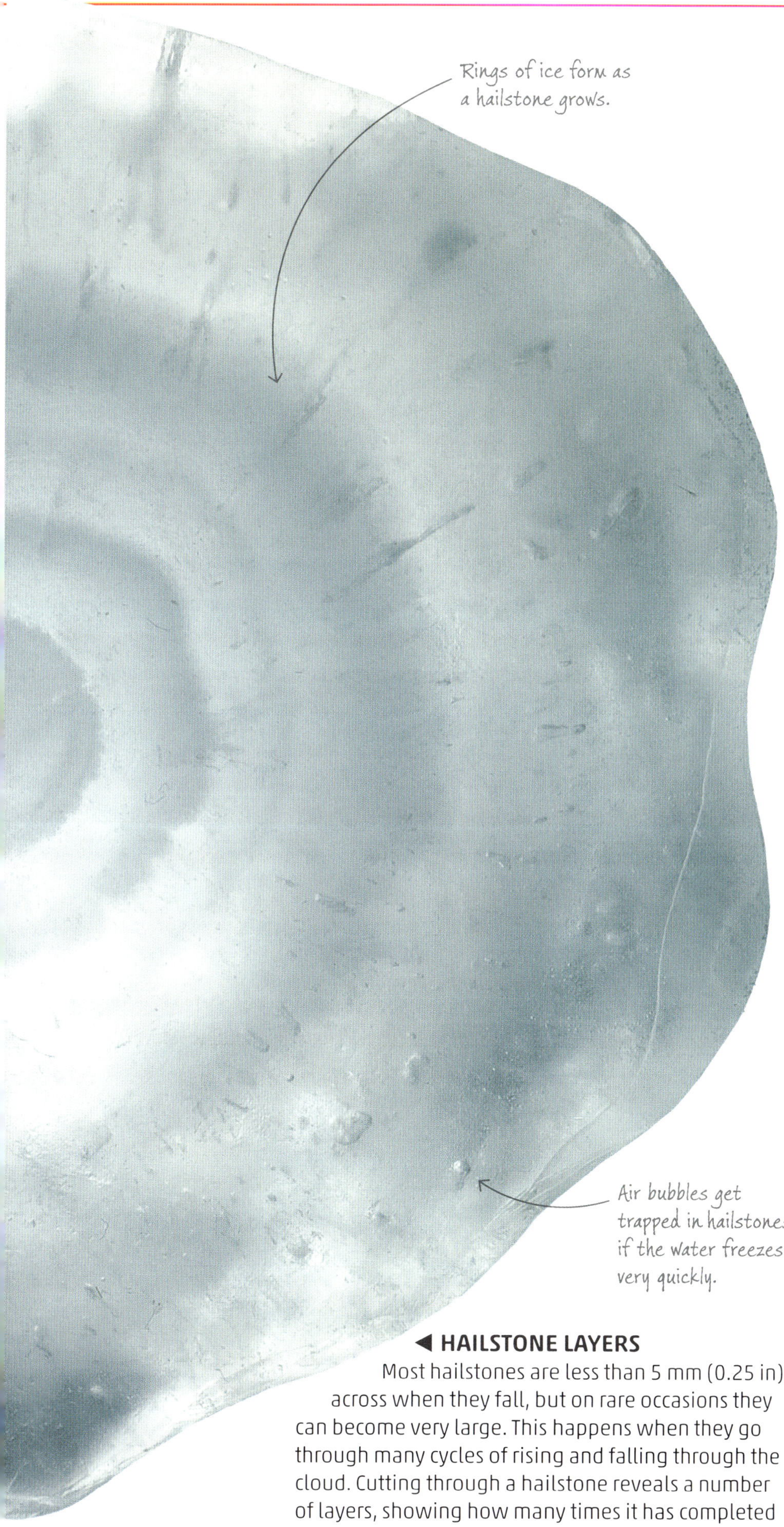

Rings of ice form as a hailstone grows.
Air bubbles get trapped in hailstones if the water freezes very quickly.

◀ HAILSTONE LAYERS

Most hailstones are less than 5 mm (0.25 in) across when they fall, but on rare occasions they can become very large. This happens when they go through many cycles of rising and falling through the cloud. Cutting through a hailstone reveals a number of layers, showing how many times it has completed a rise-and-fall cycle.

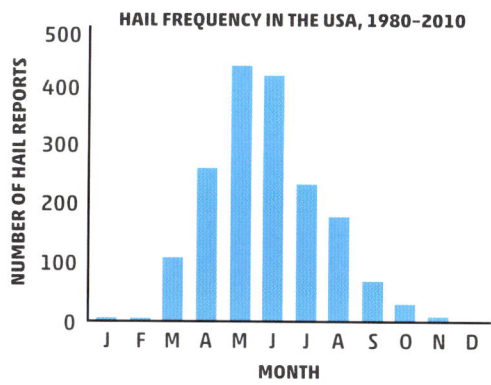

HAIL FREQUENCY IN THE USA, 1980–2010
NUMBER OF HAIL REPORTS
500
400
300
200
100
0
J F M A M J J A S O N D
MONTH

HAILSTORM SEASON

Unlike snow, hail is most common in late spring and early summer. This is when thunderstorms are common but the air temperature isn't high enough to melt the hailstones as they fall. Hailstorms can happen at any time of year in tropical countries, but they tend to occur on mountains where the air is cooler.

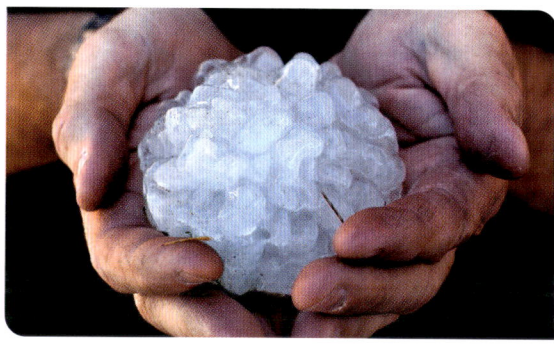

GIANT HAILSTONES

Hailstones greater than 10 cm (4 in) across are classed as "giant". The largest on record fell in North Dakota, USA. It measured 20 cm (8 in) in diameter and weighed 0.88 kg (1.9 lbs).

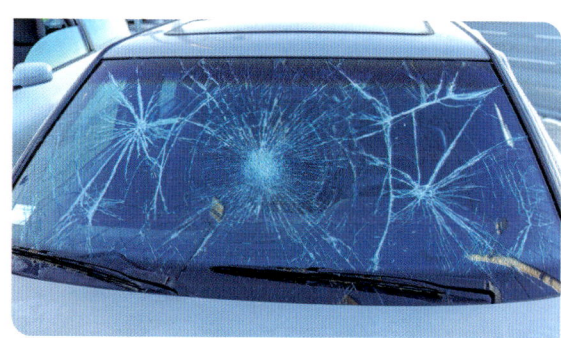

STORM DAMAGE

Large hailstones can hit the ground at 160 kph (100 mph), smashing holes in cars and houses, flattening crops, and injuring animals and people out in the open. Some hailstorms cause billions of pounds' worth of damage.

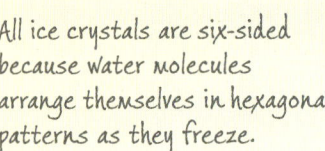

1 NEW SNOWFLAKE
A snowflake begins as a tiny speck of dust in a cloud. Water vapour from the freezing air solidifies on the speck and begins to grow into crystals.

All ice crystals are six-sided because water molecules arrange themselves in hexagonal patterns as they freeze.

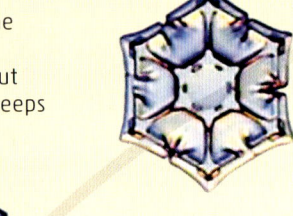
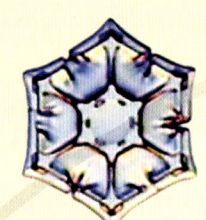

2 GROWING LARGER
Ice builds up and the snowflake grows larger. It spins as it tumbles about inside the cloud, which keeps its shape symmetrical.

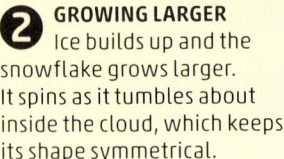

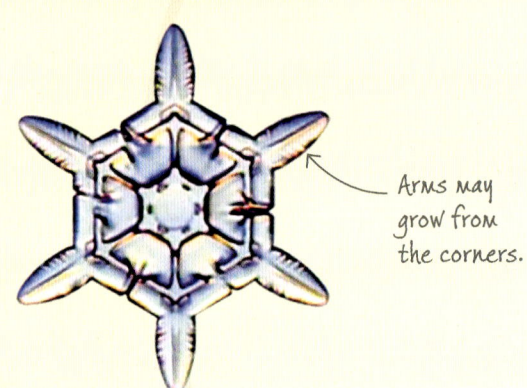

Arms may grow from the corners.

HOW SNOWFLAKES FORM

Viewed through a microscope, snowflakes reveal their amazing structure. Snowflakes are simply ice crystals, but no two are exactly alike. That's because ice crystals grow in different shapes as temperature and humidity change, and no two snowflakes experience exactly the same conditions as they tumble through a cloud. When snow finally lands it usually consists of many flakes stuck together, often half melted. However, individual snowflakes as wide as 5 cm (2 in) occasionally float to the ground intact.

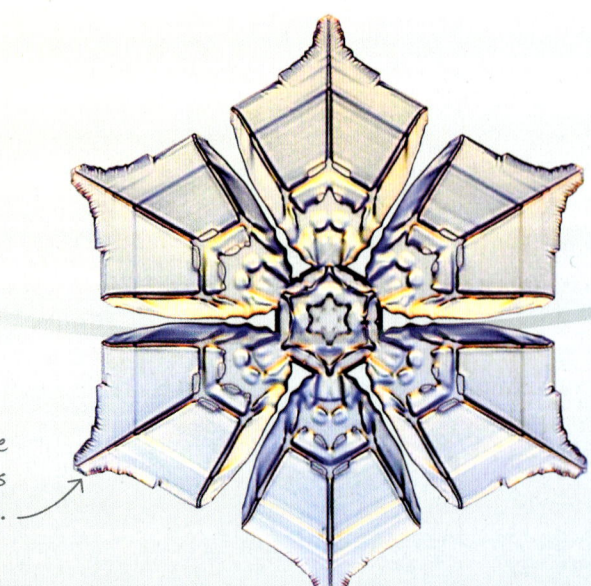

3 CHANGING CRYSTALS
Crystals continue to grow around the edge of the snowflake. As the temperature and humidity in the cloud change, the crystals take new shapes, such as arms, plates, or needles. All are based on an underlying hexagonal symmetry.

Each of the snowflake's arms grows identically.

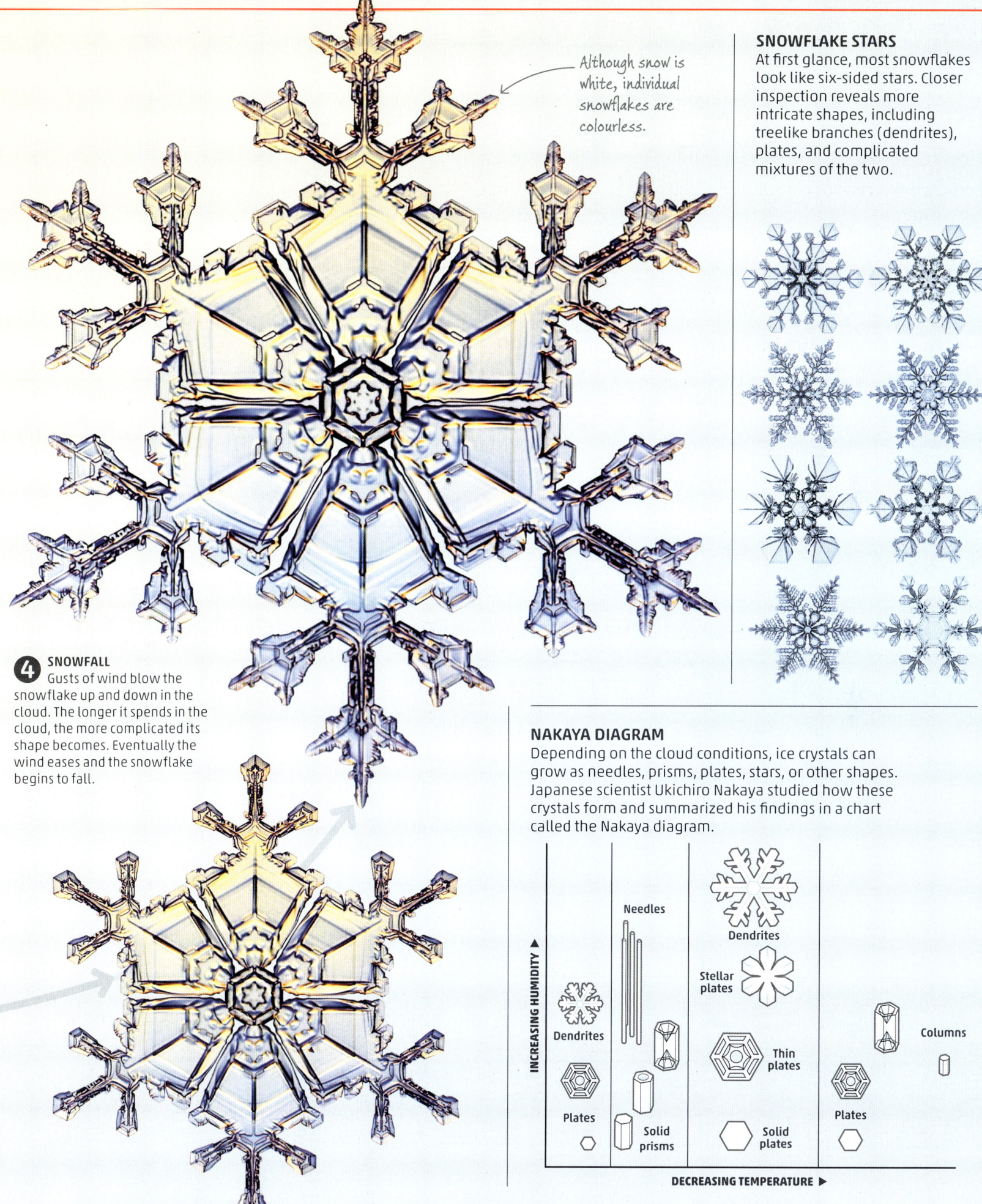

Although snow is white, individual snowflakes are colourless.

SNOWFLAKE STARS
At first glance, most snowflakes look like six-sided stars. Closer inspection reveals more intricate shapes, including treelike branches (dendrites), plates, and complicated mixtures of the two.

4 SNOWFALL
Gusts of wind blow the snowflake up and down in the cloud. The longer it spends in the cloud, the more complicated its shape becomes. Eventually the wind eases and the snowflake begins to fall.

NAKAYA DIAGRAM
Depending on the cloud conditions, ice crystals can grow as needles, prisms, plates, stars, or other shapes. Japanese scientist Ukichiro Nakaya studied how these crystals form and summarized his findings in a chart called the Nakaya diagram.

INCREASING HUMIDITY ▲

Needles

Dendrites

Stellar plates

Columns

Dendrites

Thin plates

Plates

Plates

Solid prisms

Solid plates

DECREASING TEMPERATURE ▶

TYPES OF FROST

Frost can take many different forms, from needles and feathers to a smooth layer of glassy ice.

HOAR FROST

The name of this crystalline form of frost comes from the old English word for grey. It has a feathery or sugary appearance and forms from water vapour in the air.

WINDOW FROST

Sometimes frost forms on window panes when the temperature on the outside is below freezing and there is moist, warm air on the inside. Scratches or specks of dirt act as sites for crystals to form. They grow into branching shapes like fern leaves.

GLAZE FROST

Glaze is a transparent covering of ice formed by freezing rain hitting a surface. It's known as black ice on roads as it makes the road look wet rather than icy, which can be dangerous for motorists.

AVIATION HAZARD

Rime ice is a particular hazard to planes. It adds weight, especially to the wings, and increases drag – the force of the air rubbing against the plane's surface and slowing it down. If ailerons (movable parts of the wings) ice up, it can be difficult to control the plane.

HOW FROST WORKS

Frost is a covering of ice crystals on the ground or other surfaces. It appears on clear, cold nights when the ground temperature falls below freezing. Most frost forms when water vapour in the air touches a freezing surface and turns straight into ice crystals without condensing into water first. Other kinds of frost form when dew drops freeze or when fog or mist droplets hit a freezing surface.

► RIME ICE

On windy hills and mountains, a kind of frost called rime ice sometimes appears on the windward side of rocks and trees. Rime ice forms when very cold fog droplets hit a freezing surface and turn instantly to ice.

DIRECTION OF WIND

FROST POCKETS
Cold air is heavier than warm air, which means it flows downhill and collects in valleys or hollows. This can create frost pockets – low-lying places that become covered in frost while the higher surroudings remain frost-free.

Rime ice builds up on the side of the rock facing the wind.

FROST NEEDLES
Hoar frost is made of crystals that vary in shape, depending on the temperature and humidity. At around –5°C (23°F), they grow into long needles.

Water vapour turns directly into ice when it touches the cold surface.

Rime ice often looks white and grainy.

Feathery shapes can form as rime ice builds up.

Needle-shaped ice crystals

UNUSUAL TYPES OF LIGHTNING

Some forms of lightning are seldom seen because they happen at very high altitude or are very faint, short-lived, or rare.

SPIDER LIGHTNING

These long, horizontal, spiderlike flashes on the undersides of clouds can cover vast distances as they spread from one cloud to another.

BALL LIGHTNING

In this rare form of lightning, balls of light hover over the ground for about half a minute before disappearing quietly or exploding violently.

SPRITES

Red sprites are weak discharges of electricity over active thunderclouds. Very faint and brief, they may rise 100 km (60 miles) above the cloud top.

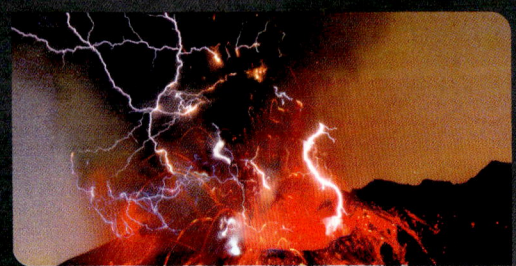

VOLCANIC LIGHTNING

Ash particles spewed out by a volcano collide and create static electricity, which generates small bolts of lightning.

HOW
LIGHTNING
WORKS

Lightning bolts are among nature's most spectacular sights. These sudden, awe-inspiring flashes of light streak from ground to sky, disappearing in a fraction of a second. A bolt of lightning is actually a giant spark of electricity in the atmosphere, usually produced by a thunderstorm.

▶ LIGHTNING STRIKE

Powerful charges of static electricity build up in a thundercloud. Eventually, this electricity is discharged as lightning, typically within the cloud or between the cloud and the ground. As lightning surges through the air, it can heat the air around it to more than 30,000°C (54,000°F), causing the air particles to glow and create an instant flash of bright light. The hot air expands so rapidly that it sends out a rumbling shock wave, which we hear as thunder. Lightning often strikes high points, such as tall buildings or – as shown here – trees.

FROM CLOUD TO GROUND TO CLOUD

As violent winds inside a thundercloud smash particles of water and ice together, the cloud becomes charged with static electricity. Positive charge builds up at the cloud's top, negative charge at its base. The negatively charged cloud base creates a positive charge on the ground, causing a two-way lightning strike from cloud to ground and back again.

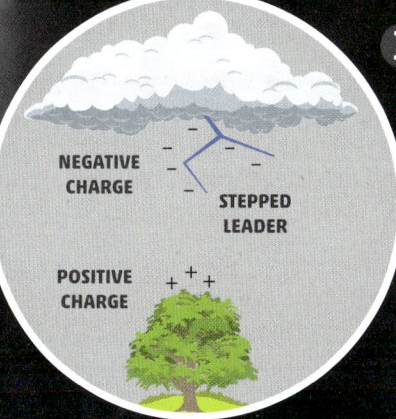

1 STEPPED LEADER

Positive charge builds up at ground level on a high point such as a treetop, tall building, or mountain peak. The opposite charges attract, pulling an electric current down from the cloud along a forked, zigzagging path. This is called a stepped leader, and it is invisible to the human eye.

NEGATIVE CHARGE

STEPPED LEADER

POSITIVE CHARGE

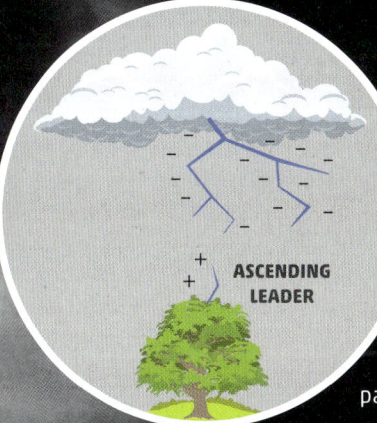

2 ASCENDING LEADER

As the stepped leader descends, it starts to draw the positive charge upwards from ground level. The rising positive charge is called an ascending leader. When the descending and ascending leaders meet, they create a conducting path through the air, made up of air molecules that have split into positive and negative particles (ions).

ASCENDING LEADER

3 RETURN STROKE

The return stroke is an electric current that passes from ground to cloud at 100,000 km per second (60,000 miles per second). This upward stroke produces the visible flash you see. The voltage of an average bolt of lightning is about a million times greater than mains electricity in a home.

RETURN STROKE

POWER SURGE

An average thunderstorm releases more energy than a nuclear bomb, much of it in the form of lightning. A bolt of lightning travels at 430,000 kph (270,000 mph) and heats the air to 30,000°C (54,000°F) – five times hotter than the surface of the Sun. The vast surge of energy tears atoms apart into charged particles and makes air expand explosively, creating a shock wave that we hear as thunder.

HOW DUSTSTORMS WORK

In some parts of the world, the weather throws up deadly duststorms that blanket everything they roll over with a layer of dirt or sand. Most duststorms occur in areas that are dry, flat, and have few trees or plants. The dust can be blown vast distances – even to the other side of the world.

▼ **DISAPPEARING IN DUST**
A duststorm and thundercloud approach the city of Yuma in Arizona, USA. The deserts of Arizona and Africa are often hit by duststorms called haboobs. These can strike without warning and move at speeds of up to 100 kph (60 mph). They reduce visibility drastically, making driving hazardous.

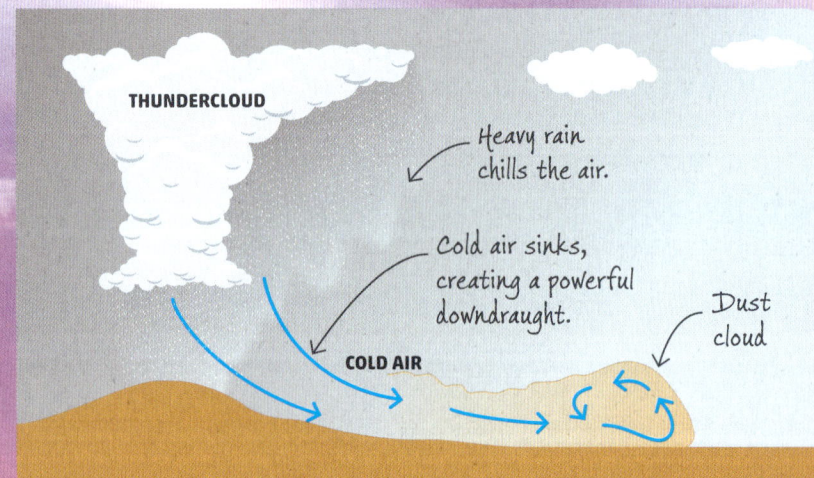

THUNDERCLOUD

Heavy rain chills the air.

Cold air sinks, creating a powerful downdraught.

Dust cloud

COLD AIR

HOW A HABOOB FORMS
A haboob is caused by a thunderstorm. When a towering thundercloud suddenly dumps its rain, the falling rain chills the surrounding air, which makes it dense and heavy. The heavy air sinks, creating a powerful wind that flows down and outwards. The wind whips up sand, dust, and other debris into a dust cloud that moves ahead of the thunderstorm. Duststorms can last for up to 30 minutes and be many miles wide.

ACROSS THE OCEAN
The wind often blows dust clouds from the Sahara over the Atlantic Ocean, engulfing the Canary Islands and turning the sky orange. Dust may even travel as far as the Americas.

DUST BOWL
In the 1930s, a series of droughts in a vast area of farmland in the USA led to crops failing. Massive duststorms followed, blowing away the topsoil and making the land – which became known as the Dust Bowl – useless for farming.

HEALTH HAZARD
Duststorms can spread diseases if fungal spores and bacteria that live in soil are inhaled. Valley fever in the southwestern USA is caused by a fungus (below) that infects the lungs. It causes breathing problems and extreme tiredness.

Cloud of dust blowing across the Atlantic

Canary Islands

SAHARA DESERT

The dust cloud may be up to 1,500 m (5,000 ft) high.

MER DE GLACE, FRANCE, IN 1910

HOW CLIMATE CHANGE WORKS

Earth's climate has changed many times in its history. There were times when the whole planet was covered in ice. At other times it was much hotter than today. Most of these changes were caused by natural processes taking place over millions of years. Today, however, the climate is changing quickly because of human activity.

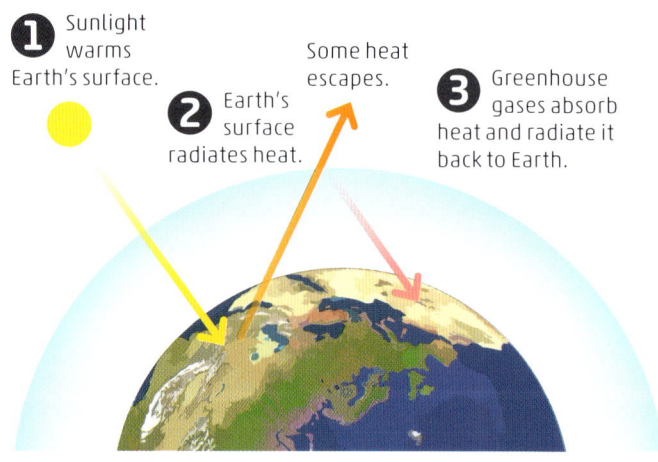

1 Sunlight warms Earth's surface.

2 Earth's surface radiates heat.

Some heat escapes.

3 Greenhouse gases absorb heat and radiate it back to Earth.

GREENHOUSE EFFECT

The main cause of climate change today is the increase in greenhouse gases, particularly carbon dioxide (CO_2) from the burning of fossil fuels, in the atmosphere. Greenhouse gases trap heat from the Sun, similar to how glass traps warmth in a greenhouse. Earth would be too cold for life without any greenhouse effect, but the increase in greenhouse gases is now making the effect too strong.

MER DE GLACE, FRANCE, IN 2012

▲ MELTING AWAY

One sign that Earth is warming is the shrinking of glaciers. For example, the Mer de Glace – a 7.5 km (4.7 mile) long valley glacier in the French Alps – lost an average of 30 cm (12 in) in height each year between 1939 and 2001. That's the equivalent loss of 280,000 Olympic-sized swimming pools full of water.

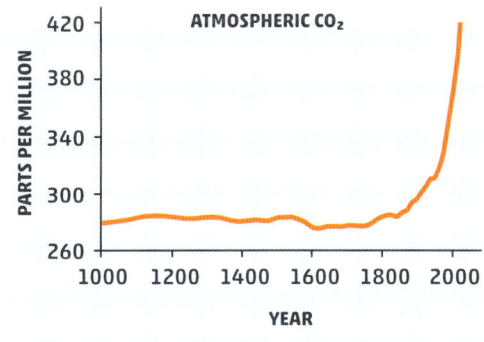

RISING CARBON DIOXIDE LEVELS

Around 200 years ago, CO_2 levels in the atmosphere started to rise steeply as people began to rely more and more on fossil fuels for energy. Deforestation to create farmland also added to CO_2 levels. Carbon stored in trees is released as CO_2 when they are felled.

STUDYING PAST CLIMATES

Scientists take deep samples of ice from glaciers in polar regions to study air bubbles trapped thousands of years ago. These studies show how CO_2 levels and climate have changed over time.

RISING METHANE LEVELS

Methane is another greenhouse gas and, like CO_2, its concentration in the atmosphere is rising. The release of methane from the thawing of permafrost in tundra regions (above) and from farming and livestock are two major causes of the increase.

The parts of Earth where life exists make up our planet's **biosphere**. Earth is the only planet known to have life. The first living things appeared at least 3.8 billion years ago, not long after oceans formed. **Microscopic** at first, they eventually **evolved** into a vast diversity of animals, plants, and other **organisms**. As life spread across the globe, it transformed the land, oceans, atmosphere, and climate.

THE BIOSPHERE

HOW **LIFE** BEGAN

The oldest evidence of life on Earth comes from fossils of single-celled microbes called cyanobacteria, which lived in the sea 3.8 billion years ago. Cyanobacteria still flourish today and have the defining feature of all life: the ability to make copies of themselves. But they weren't the first organisms on Earth. They evolved from simpler ancestors that left no trace in the fossil record. Scientists are still debating how the earliest forms of life came to exist.

Cyanobacteria thrived around coasts, where the sea is shallow, sunny, and warm.

DEEP SEA
One theory is that life began near hot springs in the darkness of the deep sea floor. The scalding water that gushes from these volcanic vents contains energy-rich nutrients that might have sustained the very first forms of life, making sunlight energy unnecessary.

► EARLY EARTH
Earth was like an alien world when the first living things appeared. There was no oxygen in the atmosphere, and the first continents were barren expanses of bare rock, scarred by craters. However, most of the planet was covered in water, which is essential to life. Life probably began in water, perhaps when a random chemical reaction happened to create a molecule that could make copies of itself.

SPARK OF LIFE
If life began in shallow rather than deep water, the first vital chemical reaction may have been triggered by the electrical energy of lightning.

Earth and the Moon were bombarded by asteroids and comets in their early years, leaving thousands of craters. Earth's craters have since disappeared, but the Moon's ancient craters remain.

MILLER-UREY EXPERIMENT

In 1953 two scientists, Stanley Miller (below) and Harold Urey, carried out an experiment to see if they could create the chemical building blocks of life from gases they thought existed in Earth's early atmosphere. They mixed water, methane, ammonia, and hydrogen in a glass flask and passed sparks through it. After a week they found amino acids, which are found in all living things today. Amino acids can't copy themselves, but similar experiments later produced the components of ribonucleic acid (RNA), a molecule similar to DNA.

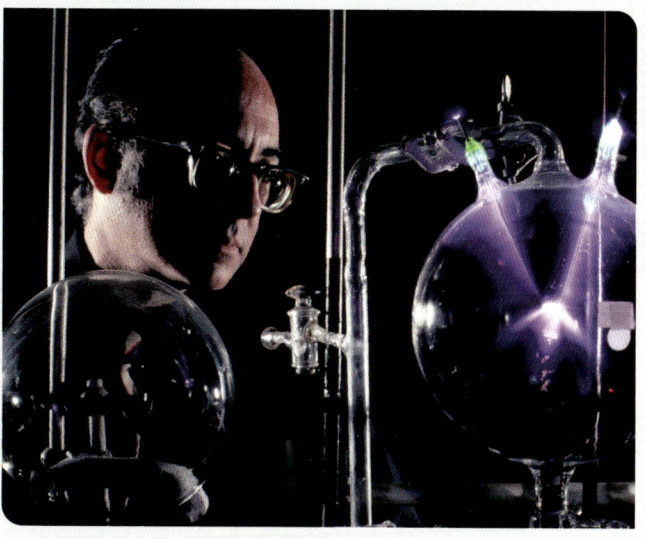

RNA WORLD

Some scientists believe the first forms of life were RNA molecules that could make copies of themselves. Like DNA, RNA can store genetic information as a kind of code. However, unlike DNA, RNA has the ability to control chemical reactions without needing proteins and other complex molecules to help it. However, the first self-copying molecule might have been something quite different that has now disappeared.

RNA is similar to DNA but is single stranded.

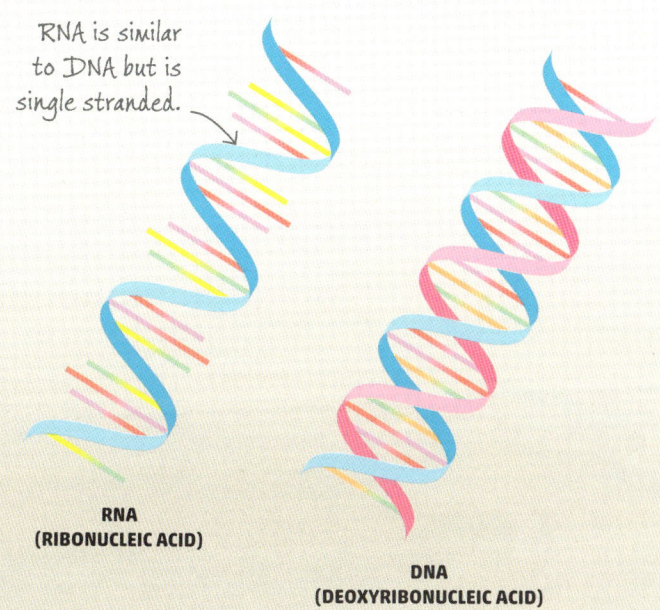

**RNA
(RIBONUCLEIC ACID)**

**DNA
(DEOXYRIBONUCLEIC ACID)**

SPACE INVADERS

A few scientists think life may have been seeded on Earth by comets. Alien microbes would have had to survive being blasted off their home planet by an impact before a long journey through space and then a second impact with Earth – which seems unlikely. But the organic molecules responsible for the origin of life might have made it.

HOW LIFE CHANGED THE
ATMOSPHERE

No animals could have survived on the early Earth because there was no oxygen for them to breathe. We owe our existence to countless billions of microscopic organisms that, over a vast span of time, produced the oxygen that made animal life possible. These organisms – cyanobacteria – were some of the first living things, and they still thrive throughout the world today.

▼ SOLAR POWER
Very early in the history of life, marine cyanobacteria developed the ability to harness solar energy and used it to make living tissue. Some formed thin mats that built up to create rocky mounds called stromatolites – just like these that still live in the warm, shallow waters of Shark Bay in Western Australia.

CYANOBACTERIA
The surface of a living stromatolite is covered in a dense mat of microscopic cyanobacteria.

Cyanobacterium

The water in this pool is unusually salty and hostile to animals that would otherwise eat the bacteria.

PHOTOSYNTHESIS

Green plants, algae, and cyanobacteria soak up solar energy and use it to turn water and carbon dioxide into sugar and oxygen. This process is called photosynthesis. They use the sugar to make more complex carbohydrates such as cellulose (plant fibre), as well as vital proteins. They release the oxygen into the surrounding water or air.

GREAT OXYGENATION EVENT

At first, most of the oxygen made by cyanobacteria combined with iron dissolved in sea water to form iron oxide, which sank to the sea floor. When the iron ran out, oxygen began to build up in the atmosphere in the "great oxygenation event". Today a fifth of air is oxygen.

LAYERS OF LIFE

A stromatolite consists of layer upon layer of cyanobacteria, each new layer having grown on the dead remains of others. Fossilized stromatolites with the same layered structure are found in rocks up to 3.5 billion years old.

This water plant's leaves are releasing bubbles of pure oxygen.

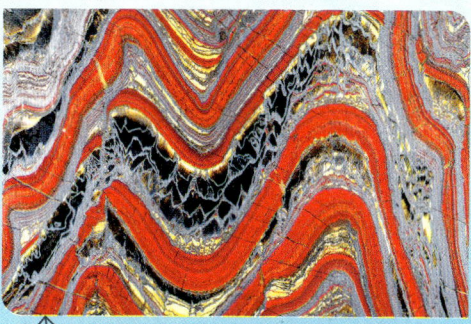

The red bands in this ancient rock formed from iron oxide in the sea more than 3 billion years ago, before Earth's air contained oxygen.

This fragment of a fossil stromatolite reveals its layered structure.

Each stromatolite is centuries old, having grown at the rate of less than 1 mm (1/25 in) a year.

1 DEATH
After the trilobite dies, the soft internal parts of its body soon decay or are nibbled away by tiny sea creatures. However, its hard external skeleton remains intact.

2 BURIAL
Particles of soft silty or muddy sediment settling on the seabed gradually bury the trilobite's remains, protecting them from further damage.

3 MINERALIZATION
As the trilobite is buried deeper, minerals dissolved in the water seep into its remains and turn them to stone. Meanwhile the soft sediments harden into rock.

TRACE FOSSILS
Some fossils preserve a record of an animal's behaviour rather than its body. Known as trace fossils, the best-known examples are footprints, like these of a predatory dinosaur. They show how it walked, ran, and even attacked its prey.

Fossilized leaves in coal

FOSSIL FUELS
When dead plants or microscopic algae decay, their carbon content turns to carbon dioxide. But if they are buried before decaying, their carbon compounds become energy-rich fossil fuels. Plants turn to coal, and algae turn to oil.

HOW
FOSSILS
FORM

Planet Earth has been home to a dazzling diversity of life forms, but most of them are now extinct. We only know that they existed because their remains are preserved in the rocks as fossils.

▼ TRILOBITE FOSSIL
Some fossils are so perfect that they almost resemble living things. Most fossils are hard-shelled sea creatures that got buried by mud before they could decay or be eaten. This trilobite fossil is about 400 million years old. Trilobites died out some 250 million years ago.

4 UPLIFT
Massive earth movements buckling Earth's crust push the rock above sea level, to become dry land. Here it is exposed to the forces of erosion.

5 EXPOSURE
Rain and wind slowly wear away the rock until the fossil is exposed. A sharp-eyed scientist spots it and carefully excavates the fossil.

The body was divided into several hard-shelled segments, like that of a scorpion.

The long spines may have been used for defence.

Curved brow horns stuck out from above its eyes.

This trilobite had an unusual fork-like horn on its head that was as long as its body.

HOW BIOMES WORK

Every living thing has a preferred habitat, which is a combination of terrain and climate. All the living things in the habitat interact in a web of life called an ecosystem. In turn, ecosystems interconnect to form distinctive geographic regions called biomes.

▶ BIOMES OF THE WORLD

The world's biomes range from barren deserts to lush rainforests, from low-lying wetlands to high mountains, and from sun-scorched tropical grasslands to icy polar tundras. Apart from wetlands, which are found in small patches all over the world, most biomes cover great swathes of land.

ARID SCRUBLANDS
Also called Mediterranean scrublands, this biome typically has hot, dry summers and cool, rainy winters. These regions have a high diversity of plants.

Key

- **TEMPERATE GRASSLANDS**
- **WETLANDS**
- **TEMPERATE FORESTS**
- **MOUNTAINS AND HIGHLANDS**
- **ARID SCRUBLANDS**
- **TROPICAL DRY FORESTS**
- **TROPICAL RAINFORESTS**
- **TUNDRAS**
- **BOREAL FORESTS**
- **DESERTS**
- **TROPICAL GRASSLANDS**

TEMPERATE FORESTS
Regions with cold winters, mild summers, and plenty of rain support forests of broadleaved trees. Many of these lose their leaves in autumn and grow new ones in spring.

TEMPERATE GRASSLANDS
Where there's not enough rain to support dense tree growth, parts of the world with temperate climates naturally develop into grasslands such as steppes and prairies.

WETLANDS
Where water floods the land it creates swamps, marshes, fens, and bogs. These wetlands are colonized by plants adapted to cope with the waterlogged ground.

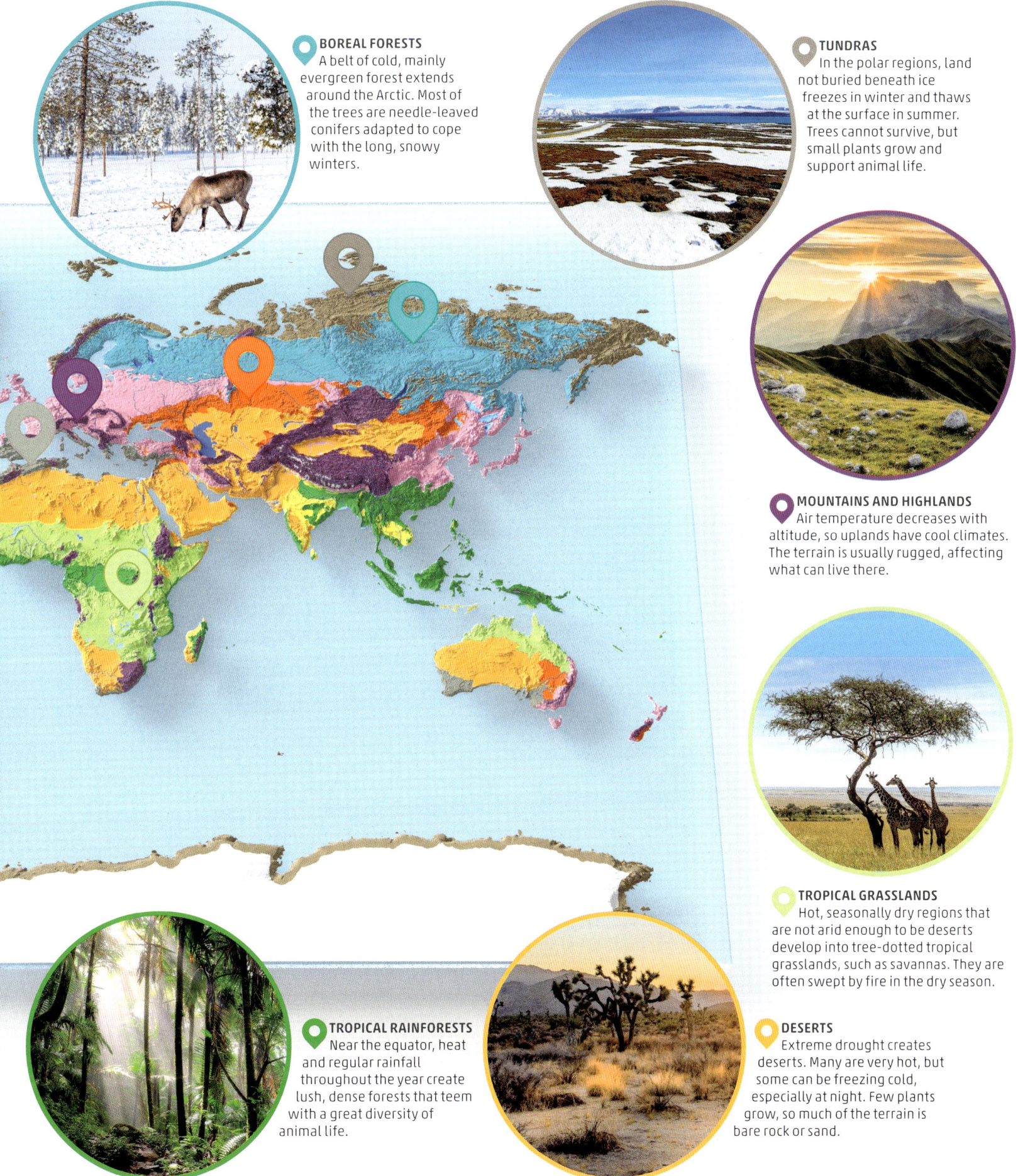

BOREAL FORESTS
A belt of cold, mainly evergreen forest extends around the Arctic. Most of the trees are needle-leaved conifers adapted to cope with the long, snowy winters.

TUNDRAS
In the polar regions, land not buried beneath ice freezes in winter and thaws at the surface in summer. Trees cannot survive, but small plants grow and support animal life.

MOUNTAINS AND HIGHLANDS
Air temperature decreases with altitude, so uplands have cool climates. The terrain is usually rugged, affecting what can live there.

TROPICAL GRASSLANDS
Hot, seasonally dry regions that are not arid enough to be deserts develop into tree-dotted tropical grasslands, such as savannas. They are often swept by fire in the dry season.

TROPICAL RAINFORESTS
Near the equator, heat and regular rainfall throughout the year create lush, dense forests that teem with a great diversity of animal life.

DESERTS
Extreme drought creates deserts. Many are very hot, but some can be freezing cold, especially at night. Few plants grow, so much of the terrain is bare rock or sand.

HOW TUNDRAS WORK

Closer to the north and south poles, it becomes harder for plants and animals to survive the cold. Trees disappear, and only low-growing plants like shrubs and grasses can thrive. This landscape is called tundra. For most of the year it's dark and covered in snow and ice. Then, for a few brief months in summer, it bursts into colour and is a frenzy of animal activity.

▼ SUMMER IN THE TUNDRA

During summer in Canada's far north, the snow melts but only the top 30 cm (12 in) or so of soil thaws out. Below this is "permafrost" – a permanently frozen layer that traps surface water, making all tundras boggy. Only small plants can thrive in this wet, shallow soil. They take advantage of the 24-hour midsummer daylight to grow quickly and flower. Swarms of mosquitoes and other insects hatch, and migratory birds arrive to raise their young.

CLOUDBERRY

Cloudberries can thrive in the wet, acidic soils of the tundra. These low-growing plants produce tangy fruits, which are eaten by mammals and birds. Their leaves are food for the caterpillars of moths and butterflies.

MOSS CAMPION

This small-leaved, evergreen plant forms a compact cushion close to the ground to protect itself from the freezing winds of the tundra.

CARIBOU

Also called reindeer, caribou are among the largest mammals on the tundra. They have a thick coat to keep warm and hooves adapted to walking in snow and scraping it away to find moss and lichens to eat.

LOCATION

The tundra forms a belt around the Arctic Ocean, with smaller areas in Antarctica and nearby islands. In recent decades, its area has been shrinking due to global warming.

CLIMATE

Tundra is one of the coldest and driest of all biomes, with less than 25 cm (10 in) of precipitation a year. In winter the temperature averages –25°C (–13°F), and in summer it rarely exceeds 15°C (59°F).

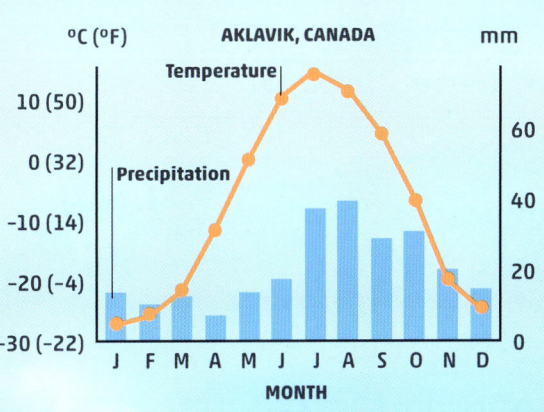

AKLAVIK, CANADA

°C (°F)   mm

Temperature

Precipitation

MONTH

POLAR BEAR

Polar bears move to the tundra in summer when the sea ice melts. They must survive without their main prey – seals – until the ice returns. Food is scarce, but they take advantage of migratory birds and their eggs, and hoofed mammals such as caribou.

LEMMING

Lemmings have thick fur and stocky bodies, which help reduce heat loss. In winter, these rodents make networks of tunnels under the snow to keep warm and to avoid predators.

SNOW GOOSE

Snow geese are migratory birds that fly to the tundra in midsummer, when the Sun never sets. This makes it easier to find food and raise their young. They like to nest near snowy owls, which deter predators such as skuas.

HOW BOREAL FORESTS WORK

A vast swathe of conifer forest encircles Earth's Arctic regions, running across Canada, Scandinavia, Russia, and Alaska. Known as boreal forest or taiga, this is the world's largest biome. The trees here have adapted to survive bitterly cold, snowy winters and short summers when the forest bursts into life.

▼ **FOREST LIFE**
The trees grow tall in the boreal forests that stretch across Canada, even though the soil freezes solid in winter. Some animals live in boreal forests throughout the year. Others migrate south in winter.

LINGONBERRY
This low-growing shrub thrives in the nutrient-poor, acidic soils of the boreal forest. It also tolerates very low temperatures in winter and the shady forest floor.

MOOSE
Moose have large feet that stop them from sinking in the snow. Their big feet are also useful for swimming when searching for aquatic plants to eat.

ATLANTIC SALMON
Boreal rivers are breeding grounds for several species of salmon. The annual rush of mature salmon upriver from the sea to spawning sites means that these fish are adapted to life in both fresh water and salt water.

LOCATION

Boreal forests are found in the northern hemisphere and lie in a wide belt around the Arctic Circle. The forests have many freshwater lakes and boggy swamps in summer.

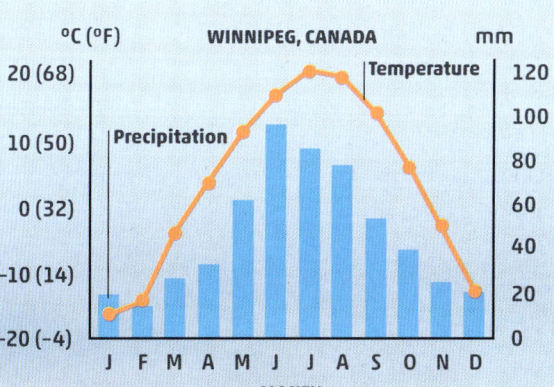

CLIMATE

There are four clear seasons in boreal forests. Winters are freezing cold, but spring and autumn are usually mild. Summer is short but warm, with very long days. It's also the rainiest time of year.

°C (°F)   WINNIPEG, CANADA   mm

Temperature

Precipitation

20 (68)   120
10 (50)   100
           80
0 (32)   60
-10 (14)   40
-20 (-4)   20
           0

J F M A M J J A S O N D

MONTH

CONIFERS

Conifer trees are conical – a shape that helps reduce the build-up of snow on their branches. Instead of flat leaves, conifers have needles to conserve water when the ground is frozen.

BOREAL CHICKADEE

This member of the tit family feeds on conifer seeds and insects living in the trees. It stores seeds for winter, when food is scarce.

WOOD FROG

Boreal forests are generally too cold for amphibians and reptiles. The wood frog, however, has a special trick – it can survive being frozen and thawed repeatedly, which would kill most other animals.

CANADIAN LYNX

This bobtailed cat has large, fur-covered paws that act like snowshoes. They spread out the cat's weight and prevent it from sinking into the snow when hunting for snowshoe hares – the lynx's main prey.

HOW TEMPERATE FORESTS WORK

In parts of the world with a temperate (mild) climate, forests flourish. The temperate forests of the northern hemisphere are dominated by broadleaved trees such as oak and maple. The trees are mostly deciduous, which means they lose their leaves in autumn. There are also evergreen temperate forests, especially in the southern hemisphere.

▼ EUROPEAN DECIDUOUS FOREST
Temperate forests have layers. The top layer is the canopy, made up of the crowns of tall trees. Below this is the understorey, where the upper branches of shorter trees reach. Next is the shrub layer, which is home to low-growing plants. The forest floor forms the bottom layer and is usually covered in twigs and decaying leaves.

GREATER SPOTTED WOODPECKER
This woodpecker is common in temperate forests and usually stays all year round. It uses its extra-strong beak to dig out beetle grubs in trees. It also stores seeds to help it survive through winter.

BLUEBELL
Bluebells are often found in old European forests. They survive winter by storing food underground in bulbs. This means they can grow quickly in spring, when the ground warms up but plenty of light reaches the forest floor while the trees are still bare.

RED DEER
These large deer feed on a wide range of plants, from grasses and shrubs to berries, acorns, and tree bark. Like most mammals living in regions with cold winters, red deer grow a thick coat to keep them warm in the chilly months.

LOCATION

Large swathes of temperate forest are found across Europe, China, and eastern North America. Evergreen temperate forests are mainly found in Chile, Australia, and New Zealand.

CLIMATE

Temperate forests grow in regions with four distinct seasons. The summers are not too hot and the winters are cool, and there is plenty of rain throughout the year.

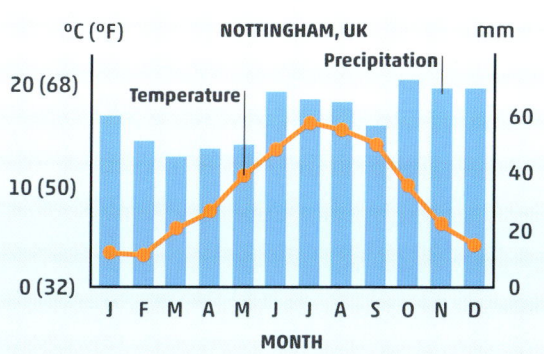

NOTTINGHAM, UK

°C (°F) — Temperature — Precipitation — mm

20 (68)
10 (50)
0 (32)

60
40
20

J F M A M J J A S O N D

MONTH

BARE BRANCHES

Broadleaved trees such as oak and beech shed their leaves each year. The trees do this to save energy and water during winter, when conditions are harsh.

RED SQUIRREL

Squirrels are the acrobats of the forest, running up and down trees and jumping from branch to branch. To survive winter, they grow a thicker coat, and they live off stores of nuts and seeds collected in autumn.

STAG BEETLE

Stag beetles lay their eggs deep underground. When they hatch, the grubs make their way to the surface to feed on dead wood and leaves, then return to the soil for winter. The adult beetles emerge in spring.

HOW TEMPERATE GRASSLANDS WORK

In parts of the world with a temperate (mild) climate, grasslands flourish in areas that have poor soil, regular wildfires, or not enough rain for forests. Grasses also dominate where there are lots of grazing animals, from cattle and sheep to bison and guanacos. These plant eaters prevent large plants from growing, but grasses grow back quickly after grazing like a lawn growing back after mowing.

PINK FAIRY ARMADILLO
This small mammal could fit in your hand. Its streamlined shape and smooth shell make it aerodynamic against the wind and also help it burrow into the sandy grassland in seconds, where it spends most of the day.

RHEA
Rheas are the largest birds in South America. They can't fly, but their long legs and necks help them spot danger, and they can outrun most predators. Rheas swallow pebbles to help them grind the tough plants that form part of their diet.

LOCATION

Temperate grasslands occur in the middle of North America and in a long swathe stretching from Eastern Europe across Central Asia. There are also areas in South America and Australia.

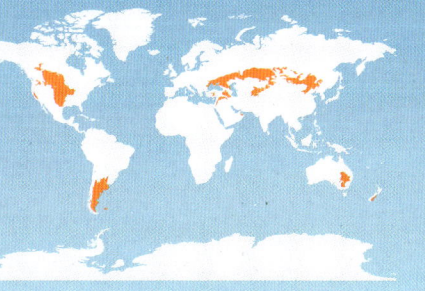

CLIMATE

Temperate grasslands have hot summers and cold winters. The lack of trees means that these places can be very windy.

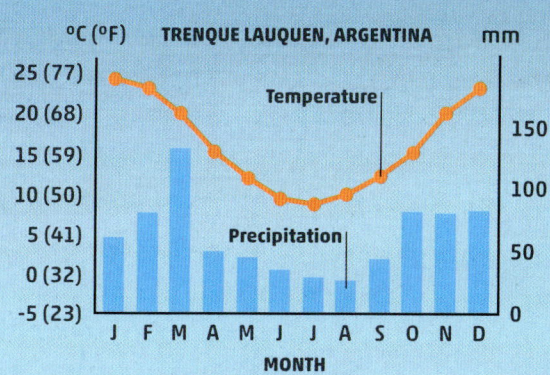

TRENQUE LAUQUEN, ARGENTINA

Temperature

Precipitation

°C (°F): 25 (77), 20 (68), 15 (59), 10 (50), 5 (41), 0 (32), -5 (23)

mm: 150, 100, 50, 0

MONTH: J F M A M J J A S O N D

GUANACO

Guanacos are ruminants – specialized grass eaters with a three-chambered stomach housing microorganisms that can break down the tough fibres in grass. Their top lip is split to help them grab and pluck grasses and leaves. They are good at conserving water and can survive in very dry places.

GIANT GRASS

One of the tallest grass species in the pampas is called simply pampas grass. It reaches 4 m (13 ft) tall and has razor-sharp grass blades to discourage herbivores. Its roots reach deep underground to find water in the dry climate.

OMBU

The ombu is the only treelike plant that grows on the pampas. It's actually a large shrub with multiple fire-resistant trunks and can survive on much less rain than a tree. Its sap and leaves are poisonous to deter animals.

MANED WOLF

This member of the dog family is a solitary animal that hunts in the cool early mornings and late evenings. Like the rhea, its long legs help it see over the tall grasses. It uses its large, sensitive ears to hunt small animals rustling through the grass.

GIANT ANTEATER

Anteaters are toothless mammals that have a long tongue to extract ants and termites from their nests. They have powerful front claws to dig for water when rivers and streams have dried up.

HOW
DESERTS
WORK

Deserts are the driest places on Earth, with less rain than any other biome. Although deserts look bare and lifeless, some plants and animals have adapted to survive with very little water. Desert plants absorb water quickly after rainfall and then store it. Desert animals avoid the hot Sun by seeking shade or by spending the day hiding in burrows.

▼ARABIAN DESERT
The Arabian Desert (below) and neighbouring Sahara cover an area larger than Europe. In these deserts, dunes made from windblown sand form vast "sand seas" with almost no plant life. In low-lying areas, however, groundwater forms salty pools surrounded by date palms and reeds. These oases provide a refuge for wildlife and human travellers.

WILD DESERT GOURD
The stems of this gourd creep along the desert sand. The plant has a deep, fleshy taproot to reach and store water so it can survive long periods without rain.

ORYX
This antelope can survive without a drink for months. The oryx gets all the water it needs from the desert plants it eats. It also produces very concentrated urine to avoid water loss.

DEATHWALKER
The highly venomous deathwalker scorpion hides in a burrow during the hot day and comes out at night to hunt. Most of its water needs are supplied by its prey – other invertebrates such as crickets and beetles.

FENNEC FOX
The fennec is a tiny desert fox. Its large ears help it get rid of heat while its hairy feet allow the fennec to walk on hot sand without burning them. Fennecs feed at night, preying on insects, small mammals, and birds.

LOCATION

Deserts are found across the world. Hot deserts, such as the Sahara, are located close to the tropics. Cold deserts, such as the Gobi, are found further north and south.

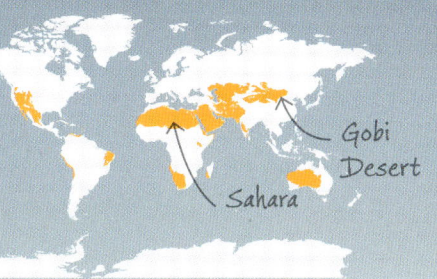

Gobi Desert

Sahara

CLIMATE

Deserts typically receive less than 25 cm (10 in) of rain each year. In some the rain may fall all at once in a storm, while in others there may not be a single drop of rain for years.

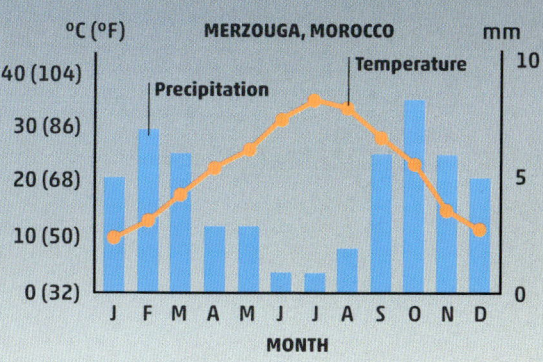

°C (°F)   MERZOUGA, MOROCCO   mm

Precipitation   Temperature

40 (104)   10

30 (86)

20 (68)   5

10 (50)

0 (32)   0

J F M A M J J A S O N D

MONTH

DESERT SPURGE

This plant may look like a cactus but it's not even a close relative. It has evolved in the same way to cope with life in a hot desert. Like a cactus, it has fleshy stems for storing water and spines instead of leaves for reducing water loss.

SAND VIPER

This venomous snake wiggles its body into the sand until just its head is poking out. It then waits to ambush passing prey. Sand vipers move across dunes in an undulating movement called sidewinding, which minimizes contact with the hot sand.

SILVER ANT

This little ant braves the desert heat for only 10 minutes a day. It has extra-long legs to keep its body clear of the hot sand. The ant is also covered in tiny silver hairs that reflect the Sun's heat.

DROMEDARY CAMEL

Dromedaries are famed for their ability to go for long periods without water. This is due to the fat stored in their hump, which breaks down to provide energy and water.

HOW TROPICAL RAINFORESTS WORK

In the wetter parts of the tropics, year-round warmth promotes the growth of dense forests. Near the equator, these get daily rain and are aptly known as tropical rainforests. They are the most biodiverse of all biomes, supporting so many plant and animal species that most of them have still not been scientifically described.

▼ BORNEO RAINFOREST
On the island of Borneo, surviving tracts of natural rainforest have a multilayered structure, with very tall emergent trees towering above an almost continuous tree canopy. Below this, smaller trees form a fragmented understorey. Most of the forest animals live in the trees, but some favour the shady forest floor.

BROMELIADS
Plants in the forest fight for light, with the tallest trees shading out the rest. Epiphytes, such as this bromeliad, have adapted to solve this problem. They live on trees high above the ground to reach sunlight.

BORNEAN ORANG-UTAN
This great ape is truly adapted for life in the trees. It uses its feet like another pair of hands when moving through the treetops. On the rare occasions when orang-utans come down to the forest floor, they find walking awkward.

TREE FROG
Sticky toe pads help tree frogs cling when climbing up and down trees. The forest is very damp and this suits the frogs, which need to keep their skin moist to absorb oxygen.

LOCATION

The main tracts of rainforest grow in tropical Central and South America, central Africa, Southeast Asia, and New Guinea.

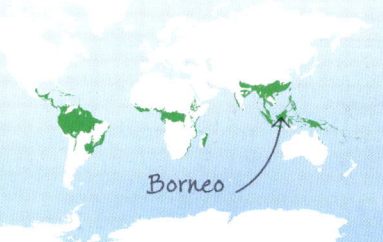

Borneo

CLIMATE

Tropical rainforests are hot and wet all year round, but in Borneo there are wetter and drier seasons. These are due to weather systems called monsoons.

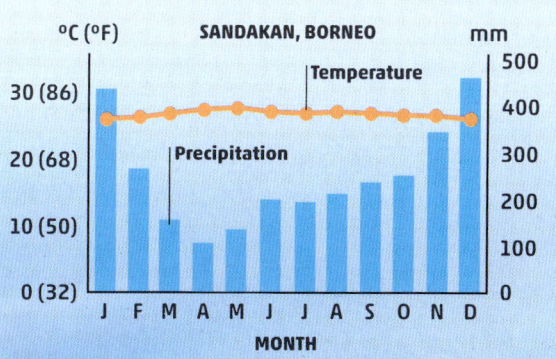

SANDAKAN, BORNEO

°C (°F)

30 (86)
20 (68)
10 (50)
0 (32)

Temperature

Precipitation

mm

500
400
300
200
100
0

J F M A M J J A S O N D

MONTH

RHINOCEROS HORNBILL

This impressive bird lives among the highest treetops, where it searches out the fruit that can be found all year round in the world's tropical rainforests.

Hornbills pluck small fruits with their beaks and skilfully toss them into their mouths.

SUN BEAR

Well equipped for forest life, the sun bear is an agile climber that spends much of its time in the trees. It eats almost anything, but especially fruit and insects.

PITCHER PLANT

Rainforest soils are poor in nutrients, so this plant traps insects and falling leaves in fluid-filled pitchers. The fluid digests the prey and extracts the nutrients the plant needs.

HOW TROPICAL GRASSLANDS WORK

Tropical grasslands (savannas) are mostly found in tropical regions that have long dry seasons that alternate with shorter rainy seasons. The harsh dry season and frequent wildfires prevent forests from growing, as do large herds of plant-eating mammals. These grazing animals are stalked by large predators such as lions and cheetahs.

MARTIAL EAGLE
As it soars high over the grasslands, a martial eagle can spot prey more than 5 km (3 miles) away. It attacks by swooping down to seize victims with its fearsome talons.

▼ THE SERENGETI
The Serengeti grassland of Tanzania in Africa still retains its original wild populations of big mammals, some of which undertake mass migrations to find food and water in the dry season. This makes the Serengeti one of the world's most spectacular wild landscapes.

UMBRELLA ACACIA
Africa's savanna is scattered with drought-tolerant trees. The umbrella acacia is named for its spreading crown. It also has very deep roots to find water and help it survive in habitats too dry for most trees.

CHEETAH
The grazing animals of the savanna rely on speed to escape predators, but few can outrun a cheetah. Able to hit 98 kph (61 mph), it is the world's fastest land animal.

LOCATION
Tropical grasslands occur in South America, India, Southeast Asia, and northern Australia, but the largest tracts are in Africa, south of the Sahara.

CLIMATE
Serengeti temperatures are generally high, but dip slightly during the long dry season. Two rainy seasons are separated by a drier period as the year begins.

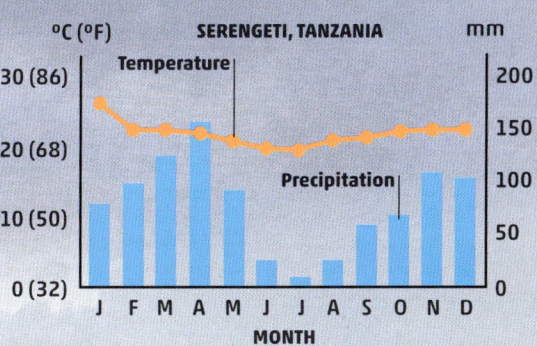

°C (°F)   SERENGETI, TANZANIA   mm
Temperature
30 (86)   200
20 (68)   150
                  Precipitation
10 (50)   100
                                 50
0 (32)   0
J F M A M J J A S O N D
MONTH

BLUE WILDEBEEST
Herds of up to 10,000 wildebeest roam the Serengeti. Like cows and sheep, wildebeest are ruminants and good at digesting grass. They are part of a mobile community of grazers that includes zebras, antelopes, and elephants.

MASAI GIRAFFE
At up to 5.5 m (18 ft) tall, the Masai giraffe is specialized for browsing the foliage of tall trees – especially the thorny acacias that dot the open savannas.

AFRICAN BAOBAB
During the rainy season a big baobab tree can absorb 120,000 litres (32,000 gallons) of water and store it in its swollen trunk. This keeps it alive in the dry season.

DUNG BEETLE
Dung beetles remove the dung of grazing animals, rolling it into balls and burying it as food for their young. Without them, the savanna would be covered in dung.

HOW MOUNTAIN LIFE WORKS

The higher you climb, the colder and windier it gets. The terrain becomes more rocky and barren, and the air gets thinner and harder to breathe. Only the toughest animals and plants can survive on mountains. On the highest peaks, life is almost impossible.

ANDEAN CONDOR
After launching itself from a cliff, this huge vulture can soar for hours on rigidly outspread wings, riding the rising air currents and scanning the ground for carrion (dead animals) to eat.

VICUÑA
The wild ancestor of the alpaca, the vicuña has the same dense woolly coat to keep it warm. Its body is specially adapted to cope with low oxygen levels at high altitude.

YARETA
Many mountain plants have a cushion-like form that resists icy winds. Yareta, an evergreen found only in the central Andes, is one of the biggest at up to 6 m (20 ft) across.

CHINCHILLA
With thicker fur than any other land animal, the chinchilla is able to live high in the mountains where temperatures often dip well below freezing at night.

LOCATION

Mountain ranges are found worldwide, mostly marking places where the moving plates of Earth's crust have been pushed together, forcing the ground upwards over millions of years.

CLIMATE

In the tropical Andes south of the equator, the average temperature is fairly stable throughout the year, but there are distinct wet and dry seasons.

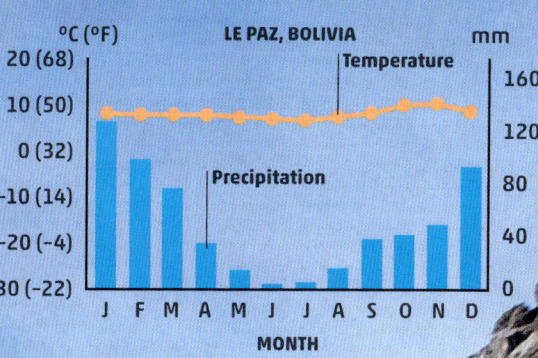

LE PAZ, BOLIVIA

°C (°F)

20 (68)
10 (50)
0 (32)
-10 (14)
-20 (-4)
-30 (-22)

mm

160
120
80
40
0

Temperature

Precipitation

J F M A M J J A S O N D

MONTH

▼ ANDES MOUNTAINS

The central Andes running through Peru and Bolivia lie near the equator, but their highest peaks – which soar to 6,000 m (20,000 ft) or more – are capped with snow throughout the year. The plants and animals living on the slopes just below the snowline are adapted to thrive in a near-Arctic climate.

LICHEN

At the highest altitudes, the only plantlike organisms are lichens. A lichen consists of a fungus containing microscopic algae, which make food by photosynthesis. This tough partnership can colonize bare rock.

JAMES'S FLAMINGO

This flamingo feeds in flocks in the mountain lakes of the central Andes, sifting the water for tiny algae with its specialized bill. Pigments in the algae turn the bird's feathers pink.

QUEEN OF THE ANDES

A giant relative of the pineapple, this spectacular plant produces a flower spike up to 8 m (26 ft) tall. Its flowers are pollinated by birds.

ANDEAN MOUNTAIN CAT

The size of a large domestic cat, this thick-furred hunter lives in rocky terrain, where it preys mainly on the rabbit-like mountain viscacha.

HOW ARID SCRUBLANDS WORK

In parts of the world with a Mediterranean-type climate, the plant life is dominated by low-growing, woody shrubs. These have tough, leathery leaves adapted to reduce water loss during the hot, dry summers. Many can even survive the regular wildfires that sweep over the landscape. Big animals are rare, but many smaller, specialized animals thrive.

▼ THE FYNBOS

The fynbos on the southern tip of Africa is a region of astonishing biodiversity, with more than 8,500 species of plants growing in dazzling profusion. Most of these grow nowhere else in the world. Their colourful flowers attract nectar-feeding, pollinating insects and birds, while small mammals and reptiles forage for food on the ground below.

ORANGE-BREASTED SUNBIRD

Like American hummingbirds, sunbirds are glittering, iridescent nectar-feeders. The orange-breasted sunbird lives only in the fynbos, where it is a pollinator of proteas and other plants.

KING PROTEA
Proteas are among the most spectacular flowers of the fynbos. Like the fire lily, they can cope with long droughts and have a thick underground stem that regenerates the plant after wildfires.

ROCK HYRAX
A rabbit-sized relative of the elephant, the rock hyrax favours the rockier parts of the fynbos, where it gathers in large groups to feed on plants, small lizards, and insects.

LOCATION

Arid scrublands occur in California, Chile, parts of southern Australia, on the southern tip of South Africa, and all around the Mediterranean.

CLIMATE

Like other regions of arid scrubland throughout the world, the South African fynbos has mild, damp winters and hot, very dry summers (November to March).

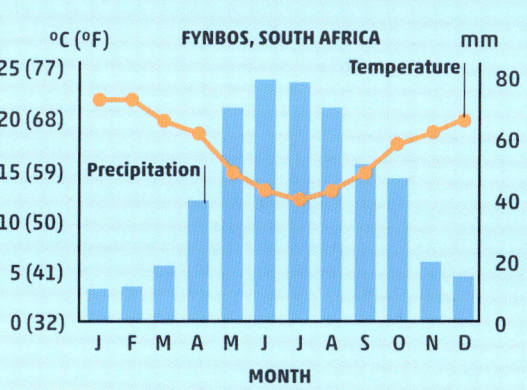

FYNBOS, SOUTH AFRICA

°C (°F) / mm

Temperature

Precipitation

25 (77), 20 (68), 15 (59), 10 (50), 5 (41), 0 (32)

80, 60, 40, 20, 0

MONTH: J F M A M J J A S O N D

FIRE LILY

Vivid red fire lilies live up to their name by sprouting from hidden bulbs after summer wildfires. The plant reacts to smoke, blooming within two weeks of a fire.

ROYAL HEATH

Heathers are common scrubland plants. The royal heath is one of the most colourful, found only in the most southerly parts of the fynbos, where it flowers all through the year.

TABLE MOUNTAIN BEAUTY

This distinctive butterfly is the sole pollinator of several fynbos flowers, including the fire lily. The females are slightly larger than the males, with a wingspan of up to 9 cm (3.5 in).

GEOMETRIC TORTOISE

Now very rare, this small, strikingly marked tortoise may spend weeks hiding away from the hot summer sunshine. When active in cooler weather, it eats grasses and plant bulbs.

CAPE GRYSBOK

This small antelope hides in the shade by day and emerges at night to look for food. It lives alone, and rival males use their short, sharp horns to fight fiercely over territory.

HOW WETLANDS WORK

Broad expanses of shallow water gradually become choked with aquatic plants, creating swamps, marshes, mangroves, fens, and bogs. Some of these wetlands extend over huge areas, expanding and contracting with the seasons. They provide safe refuges for all kinds of animals, but especially those that are adapted to live, feed, or hunt in water or on soggy ground.

▼ THE PANTANAL

The Pantanal in central South America is the world's largest tropical wetland as well as the world's largest freshwater wetland. It lies in a shallow depression fed by rivers flowing off the surrounding uplands. The rivers bring water and sediment that settles to create a rich mosaic of pools and marshes teeming with spectacular wildlife.

ROSEATE SPOONBILL
Wading slowly through the marshes and shallow pools, the roseate spoonbill swings its specialized beak from side to side to sift small animals from the water.

CAPYBARA
The capybara is a giant rodent that can weigh as much as an adult human. It lives in small, semi-aquatic herds that wade in the shallows to feed on water plants.

YELLOW ANACONDA
One of the world's longest, heaviest snakes, the yellow anaconda is an excellent swimmer that hunts in shallow water for prey, including capybaras and small deer.

SALT MARSHES

Coastal wetlands affected by tides have salt water rather than fresh water. In countries with a cool climate, salt marshes flourish. They provide important refuges for migrating birds and many kinds of waterfowl.

MANGROVE FORESTS

Mangrove forests are found in the tidal zone of tropical coasts. They are dominated by salt-tolerant trees with stiltlike roots for support in soft mud. This sheltered habitat is an important nursery for ocean fish and other marine animals.

GIANT WATERLILY

The floating leaves of this water plant reach 2 m (7 ft) wide. They are often visited by foraging waders such as the wattled jacana – the bird's long toes spread its weight so it can walk on even the flimsiest floating plants.

WATER HYACINTH

This free-floating plant multiplies rapidly to form mats that cover large areas of water. The plant is kept afloat by buoyant, bulb-like nodules at its base.

BLADDERWORT

The innocent-looking bladderwort preys on small aquatic animals, trapping them in purse-like bladders on its roots and digesting them to absorb nutrients.

YACARE CAIMAN

This semi-aquatic cousin of the American alligator has a varied diet that includes snails, snakes, and fish – especially piranhas. It gets through about 40 sets of teeth during its lifetime.

JAGUAR

Jaguars prowl through the wooded margins of the marshes, and even plunge into the water to seize prey. The Pantanal is one of the strongholds of this powerful hunter.

HOW
OCEAN LIFE
WORKS

Oceans cover more than two-thirds of the planet. They contain a diverse range of habitats, from icy polar seas to tropical coral reefs, and from glittering surface waters to the black depths of the ocean floor. But most oceanic life lives near the sunlit surface, where tiny plankton use the energy of sunlight to make food.

▼ **CORAL REEFS**
Sometimes called the rainforests of the ocean, coral reefs – such as this one in the Indian Ocean – provide a home for at least a quarter of the world's marine species. The corals that build reefs rely on microscopic algae in their bodies to harness the energy of sunlight and make food. Rising ocean temperatures make corals eject the algae, which can cause reefs to die.

Corals are filter-feeding invertebrates that live in colonies.

The hawksbill sea turtle lives in coral reefs and feeds mostly on sponges.

OCEAN ZONES

Scientists who study oceans divide the water into zones based on their depth. As you descend from the surface, the water becomes colder and darker, and pressure increases. The animals living in each zone are adapted to its conditions.

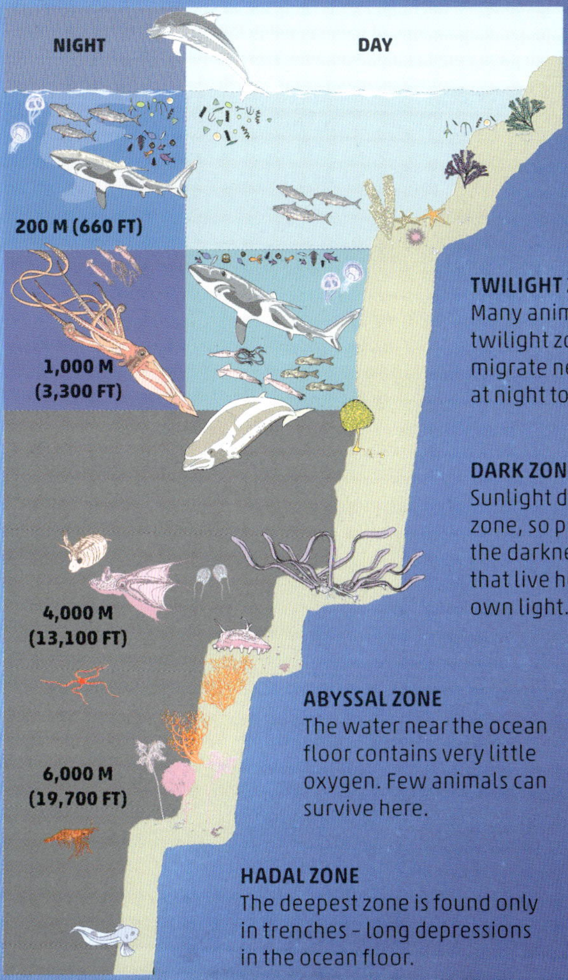

NIGHT

DAY

200 M (660 FT)

1,000 M (3,300 FT)

4,000 M (13,100 FT)

6,000 M (19,700 FT)

SUNLIT ZONE
Microscopic algae drifting in this zone use photosynthesis to make food.

TWILIGHT ZONE
Many animals hide in the twilight zone by day but migrate nearer the surface at night to feed.

DARK ZONE
Sunlight doesn't reach this zone, so prey is hard to see in the darkness. Some animals that live here produce their own light.

ABYSSAL ZONE
The water near the ocean floor contains very little oxygen. Few animals can survive here.

HADAL ZONE
The deepest zone is found only in trenches – long depressions in the ocean floor.

TIDAL SHORES
The shallow fringes of the oceans are among the most diverse marine habitats. Many of the animals are adapted to survive the tides that leave them high and dry for part of each day.

POLAR OCEANS
The cold oceans around the North Pole and Antarctica teem with life. Upwelling ocean currents bring nutrients to the tiny plankton living in the surface waters. This attracts a host of plankton feeders, which in turn attract many predators, including penguins.

DEEP SEA
From 1,000 m (3,300 ft) down to the ocean floor, prey is hard to find in the pitch-black depths. Anglerfish attract it with luminous lures, while gulper eels have huge, elastic stomachs that enable them to swallow any animal they encounter.

HUMAN IMPACT

As the world's population has grown, the area of land used for agriculture has soared, transforming our planet's appearance. About a third of Earth's total land area is now used for agriculture. Even areas once considered too dry for crops can be made fertile. Here, a circular field of corn is grown in the Mexican desert, watered by a giant rotating sprinkler.

This reference section tells you more about Earth's deep **history**, its spectacular **landscapes**, and the incredible diversity of **rocks**, **minerals**, **fossils**, and **gemstones** that exist on our planet.

▶ THE MODERN TIMESCALE

Earth's geological calendar is divided into four very long spans known as eons, a bit like the seasons on a normal calendar. These are divided into eras, each made up of one or more geological periods. The names of periods reflect the major sedimentary rock strata studied by geologists.

We are living in the Phanerozoic. Before this eon, life existed but was not widespread across the globe.

EON		ARCHEAN				PROTEROZOIC		
						PRECAMBRIAN		
ERA	HADEAN	EOARCHEAN	PALEOARCHEAN	MESOARCHEAN	NEOARCHEAN	PALEOPROTEROZOIC	MESOPROTEROZOIC	NEOPROTEROZOIC
PERIOD						Siderian / Rhyacian / Orosirian / Statherian	Calymmian / Ectasian / Stenian	Tonian / Cryogenian / Ediacaran
MILLION YEARS AGO	4,600	4,000	3,600	3,200	2,800	2,500 · 2,300 · 2,050 · 1,800	1,600 · 1,400 · 1,200	1,000 · 720 · 635 · 542

EON	PHANEROZOIC			
ERA	PALEOZOIC		MESOZOIC	
PERIOD	Carboniferous	Permian	Triassic	Jurassic

EPOCH / AGE (Carboniferous): Mississippian (Lower – Tournaisian; Middle – Visean; Upper – Serpukhovian), Pennsylvanian (Lower – Bashkirian; Middle – Moscovian; Upper – Kasimovian, Gzhelian)

EPOCH / AGE (Permian): Cisuralian (Asselian, Sakmarian, Artinskian, Kungurian), Guadalupian (Roadian, Wordian, Capitanian), Lopingian (Wuchiapingian, Changhsingian)

EPOCH / AGE (Triassic): Lower (Induan, Olenekian), Middle (Anisian, Ladinian), Upper (Carnian, Norian, Rhaetian)

EPOCH / AGE (Jurassic): Lower (Hettangian, Sinemurian, Pliensbachian, Toarcian), Middle (Aalenian, Bajocian, Bathonian, Callovian), Upper (Oxfordian, Kimmeridgian)

MILLION YEARS AGO: 358.9 · 345.3 · 328.3 · 318.1 · 311.7 · 307.2 · 303.4 · 299.0 · 294.6 · 284.4 · 275.6 · 270.6 · 268.0 · 265.8 · 260.4 · 253.8 · 251.0 · 249.5 · 245.9 · 237.0 · 228.7 · 216.5 · 203.6 · 199.6 · 196.5 · 189.6 · 183.0 · 175.6 · 171.6 · 167.7 · 164.7 · 161.2 · 155.6 · 150.8

THE GEOLOGICAL TIMESCALE

Our planet is around 4.5 billion years old. This is a mind-boggling number – just counting to 4.5 billion would take more than 145 years. In our own species' short time on Earth we have used the most solid clues available – rocks – to piece together a calendar of the planet's entire history.

INDEX FOSSILS

Fossils were the first clues used to work out the geological timescale. Deeper layers are usually older, though layers are sometimes folded or overturned. Comparing fossils found in sedimentary rocks is a better way to work out their age.

- Gastropod
- Trilobite
- Crinoid
- Coral
- Fern
- Ammonite

Matching fossils show these rock strata are the same age.

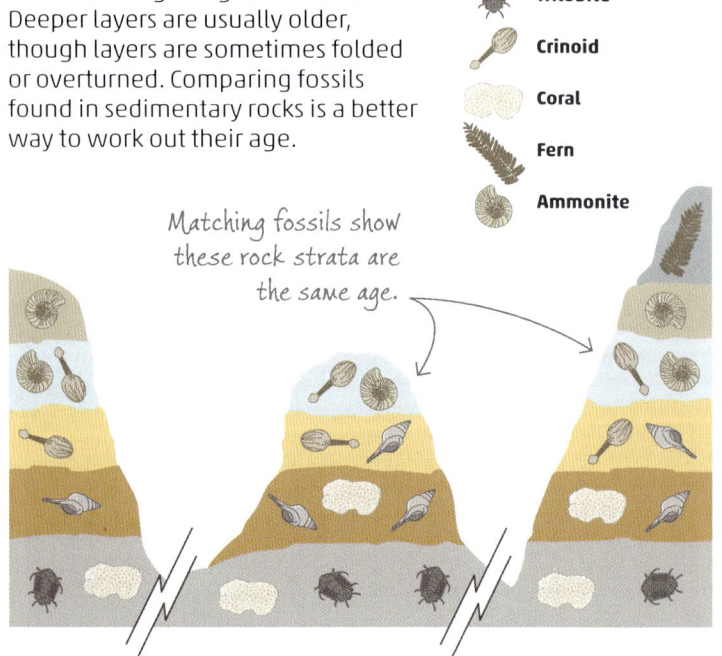

PHANEROZOIC — PALEOZOIC

	Cambrian				Ordovician			Silurian				Devonian			
PERIOD	Cambrian				Ordovician			Silurian				Devonian			
EPOCH	Lower Series	Series 2	Series 3	Furongian	Lower	Middle	Upper	Llandovery	Wenlock	Ludlow	Pridoli	Lower	Middle	Upper	
AGE	Fortunian / Stage 2 / Stage 3 / Stage 4	Wuliuan / Drumian / Guzhangian	Paibian / Jiangshanian / Stage 10		Tremadocian / Floian	Dapingian / Darriwilian / Sandbian	Katian / Hirnantian	Rhuddanian / Aeronian / Telychian	Sheinwoodian / Homerian	Gorstian / Ludfordian		Lochkovian / Pragian / Emsian	Eifelian / Givetian	Frasnian / Famennian	

EON: PHANEROZOIC **ERA:** PALEOZOIC

MILLION YEARS AGO: 539, 529, 521, 514, 509, 504.5, 500, 497, 494, 489.5, 486.4, 477.7, 470, 467.3, 458.4, 453, 445.2, 443.8, 440.8, 438.5, 433.4, 430.5, 427.4, 425.6, 423, 419.2, 410.8, 407.6, 393.3, 387.7, 382.7, 372.2, 358.9

PHANEROZOIC — MESOZOIC / CENOZOIC

EON: PHANEROZOIC

ERA: MESOZOIC | CENOZOIC

PERIOD: Cretaceous | Paleogene | Neogene | Quaternary

EPOCH:
- Cretaceous: Lower, Upper
- Paleogene: Paleocene, Eocene, Oligocene
- Neogene: Miocene, Pliocene
- Quaternary: Pleistocene, Holocene

AGE:
- (Jurassic) Tithonian
- Cretaceous: Berriasian, Valanginian, Hauterivian, Barremian, Aptian, Albian, Cenomanian, Turonian, Coniacian, Santonian, Campanian, Maastrichtian
- Paleogene: Danian, Selandian, Thanetian, Ypresian, Lutetian, Bartonian, Priabonian, Rupelian, Chattian
- Neogene: Aquitanian, Burdigalian, Langhian, Serravallian, Tortonian, Messinian, Zanclean, Piacenzian
- Quaternary: Gelasian, Calabrian, Ionian, Upper

MILLION YEARS AGO: 145.5, 140.2, 133.9, 130.0, 125.0, 112.0, 99.6, 93.6, 88.6, 85.8, 83.5, 70.6, 65.5, 61.1, 58.7, 55.8, 48.6, 40.4, 37.2, 33.9, 28.4, 23.03, 20.43, 15.97, 13.82, 11.608, 7.246, 5.332, 3.600, 2.588, 1.806, 0.781, 0.126, 0.0117

RADIOMETRIC DATING

Some minerals contain atoms that are unstable and break apart (decay) over time, changing into different kinds of atom. For example, certain uranium atoms decay into atoms of lead. This happens at a steady, predictable rate. By comparing the amounts of each atom, geologists can calculate how old a rock is.

Some zircon crystals have been dated to billions of years ago.

MASS EXTINCTIONS

The ends of some of the major divisions in Earth's history are marked by mass extinctions, when many species disappeared. One of the biggest was at the end of the Permian period. Another one wiped out dinosaurs at the end of the Cretaceous.

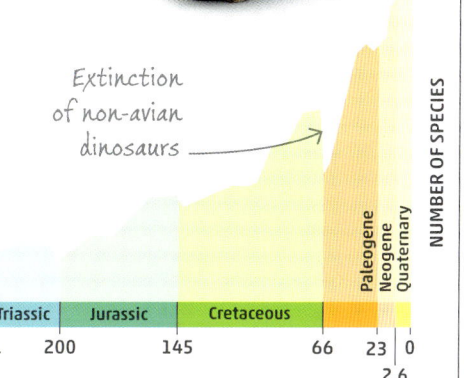

Extinction of non-avian dinosaurs

NUMBER OF SPECIES

Cambrian	Ordovician	Silurian	Devonian	Carboniferous	Permian	Triassic	Jurassic	Cretaceous	Paleogene / Neogene / Quaternary
539	486	444	419	359	299	251	200	145	66 / 23 / 0

2.6

MILLION YEARS AGO

THE FIRST GEOLOGICAL MAP

The first modern geological map of an entire country was compiled by the English geologist and surveyor William Smith in 1815. Smith pioneered the use of fossils to establish the ages of strata.

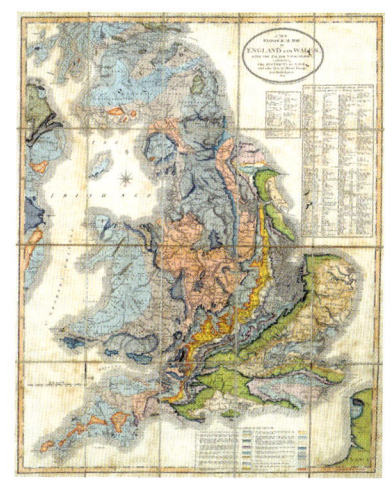

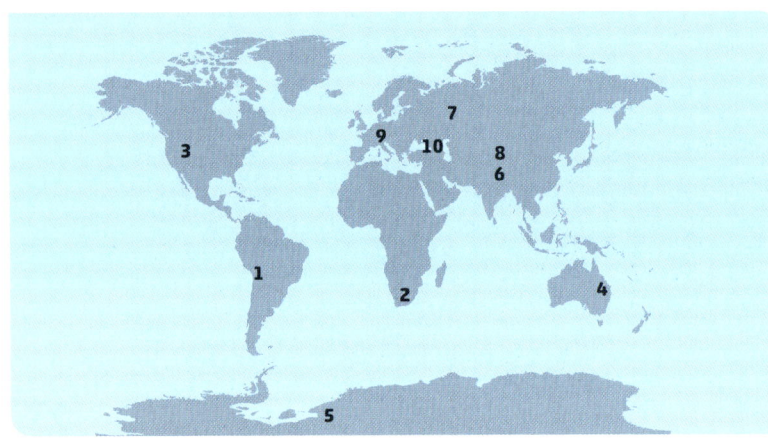

TOP TEN LONGEST
MOUNTAIN
RANGES

Mountain ranges form where two tectonic plates converge, crumpling Earth's crust and raising the land to create some of the most distinctive and awe-inspiring landscapes on Earth. A mountain is defined as a landform that's over 300 m (1,000 ft) in height, with a small number reaching almost thirty times that to where the air is so thin that mountaineers need bottled oxygen to survive.

▼ 2. SOUTHERN GREAT ESCARPMENT, SOUTHERN AFRICA
5,000 km (3,100 miles)

Located mainly in South Africa and Lesotho, the Southern Great Escarpment crosses into Zimbabwe, Namibia, and Angola. An escarpment is where there is a steep change in the height of the terrain.

▼ 3. ROCKY MOUNTAINS, NORTH AMERICA
4,800 km (3,000 miles)

The Rockies extend from New Mexico in the southwestern USA up into northern Canada. They're made up of around 100 ranges that are grouped into Canadian, Northern, Middle, and Southern Rockies.

1. ANDES MOUNTAINS, SOUTH AMERICA ▶
8,900 km (5,500 miles)

The Andes stretch down the western side of South America, from Venezuela in the north to the south of Argentina. The varied scenery includes glaciers, volcanoes, deserts, salt flats, grasslands, and rainforests. The highest peak is Argentina's Mounta Aconcagua at 6,959 m (22,831 ft).

◀ 8. TIEN SHAN, ASIA
2,400 km (1,490 miles)

Straddling the border between China and Kyrgyzstan and terminating in the southern plains of Kazakhstan is the Tien Shan range. It varies in width from 500 km (300 miles) to 350 km (220 miles) and rises to 7,439 m (24,406 ft) at the top of Victory Peak.

▲ 4. GREAT DIVIDING RANGE, AUSTRALIA
3,700 km (2,300 miles)

The Great Dividing Range is the longest range within a single country. It runs down the eastern side of Australia from Queensland to Victoria, staying almost parallel to the coast. The mountains are highest in the south, peaking with Mount Kosciuszko at 2,228 m (7,310 ft).

▲ 7. URAL MOUNTAINS, RUSSIA
2,500 km (1,550 miles)

Located in western Russia, the Urals form a natural boundary between Europe and Asia. The range extends from the Kara Sea at its northern end down to Kazakhstan, changing from frigid Arctic tundra, through forests, and into semi-deserts. Mount Narodnaya marks the highest point in the Urals at 1,895 m (6,217 ft).

▲ 9. ALPS, EUROPE
1,200 km (750 miles)

This range runs from southern France and forms a crescent shape that crosses eleven countries. Mont Blanc is the highest point at 4,807 m (15,771 ft), with the Matterhorn (above) and the Eiger being among the most recognizable peaks in the world.

▲ 5. TRANSANTARCTIC MOUNTAINS, ANTARCTICA
3,200 km (2,000 miles)

Much of this vast mountain range is buried under ice, with only the peaks showing. The range divides Antarctica into east and west regions. Mount Kirkpatrick is the highest peak in the range at 4,528 m (14,855 ft).

▲ 6. HIMALAYAS, ASIA
2,600 km (1,600 miles)

Stretching from Pakistan in the west to Bhutan in the east,, the Himalayas includes 9 of the 10 highest peaks in the world, including Qomolangma Feng (Mount Everest), which stands at 8,848.86 m (29,031 ft).

◀ 10. GREATER CAUCASUS, ASIA
Length: 1,200 km (750 miles)

The Greater Caucasus range runs from the Black Sea to the Caspian Sea, passing through Russia, Georgia, and Azerbaijan, with the Lesser Caucasus located to the south. Mount Elbrus marks the highest point, reaching 5,642 m (18,510 ft).

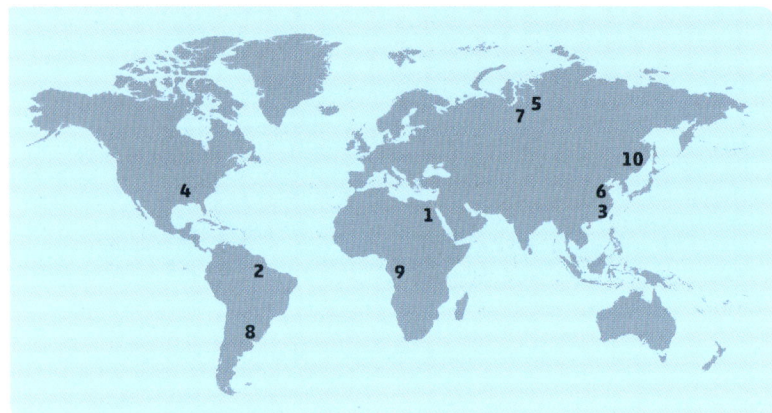

TOP TEN
LONGEST
RIVERS

Starting high on hills, mountains, and plateaus before surging across the land and draining into the oceans, rivers are vital to many aspects of life, and can have a dramatic effect on the shape of the landscape.

▲ 3. CHANG JIANG, CHINA
6,300 km (3,915 miles)

As well as being the third longest in the world, Chang Jiang is also the longest river that runs within a single country. It starts in Qing Zang Gao Yuan and flows to the East China Sea, mainly through mountainous terrain.

▼ 4. MISSISSIPPI-MISSOURI RIVER, USA
5,971 km (3,710 miles)

Starting in Lake Itasca in Minnesota, the Mississippi crosses ten states from north to south before draining into the Gulf of Mexico via a huge delta to the southeast of New Orleans. The Missouri river flows into the Mississippi, and so they are thought of as one river system, which makes the Mississippi-Missouri the fourth-longest river.

▲ 1. NILE RIVER, AFRICA
6,650 km (4,132 miles)

Widely recognized as the world's longest river, the Nile flows through eleven African nations before reaching the Mediterranean Sea.

2. AMAZON RIVER, SOUTH AMERICA ▶
6,400 km (4,000 miles)

The Amazon originates in the Peruvian Andes before flowing east through the Amazon rainforest to the Atlantic Ocean on the coast of Brazil. Although not the world's longest river, the Amazon has the largest drainage basin and greatest discharge volume of all the world's rivers.

◄ 5. YENISEI-ANGARA-SELENGA RIVER, MONGOLIA AND RUSSIA
5,539 km (3,442 miles)

The Yenisei system runs north from the Russian republic of Tyva, making its way up through central Siberia before emerging into the Arctic Sea. Snow is the main source of water, with the remainder coming from rainfall and groundwater.

▼ 6. HUANG HE, CHINA
5,464 km (3,395 miles)

Huang He starts in Qing Zang Gao Yuan and flows northeast to the Bo Hai Sea. It's the most sediment-filled river in the world due to crossing land that has easily eroded soil.

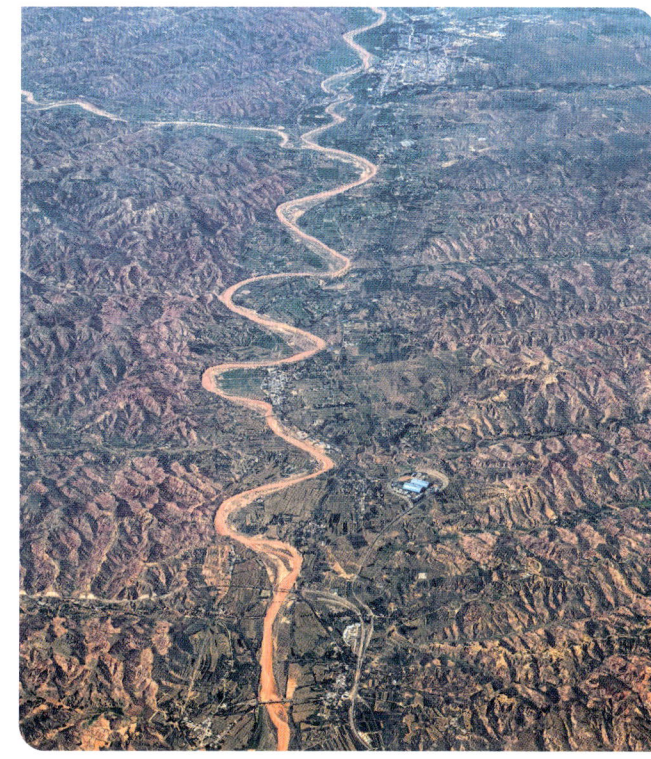

► 7. OB RIVER, RUSSIA
5,410 km (3,362 miles)

The Ob originates in the Altai mountains. It's a major transportation route, also used for irrigation and hydroelectricity.

◄ 8. PARANÁ RIVER, SOUTH AMERICA
4,880 km (3,032 miles)

The second-longest river in South America after the Amazon, the Paraná runs from southeast central Brazil, through Paraguay, and on to northern Argentina. It joins with the Uruguay River to form a vast, muddy estuary named the Río de la Plata.

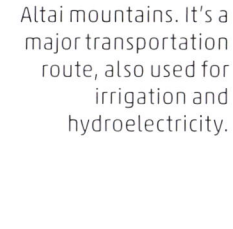

IGUAÇU FALLS ON THE PARANÁ RIVER

◄ 9. CONGO RIVER, WEST AFRICA
4,700 km (2,920 miles)

With a depth in places of 220 m (720 ft), the Congo is the deepest river in the world. It passes through nine countries, starting in northeastern Zambia and ending in the Atlantic Ocean.

▲ 10. AMUR RIVER, ASIA
4,444 km (2,761 miles)

Forming a natural border between eastern Russia and northern China, the Amur originates in northeastern Mongolia near the sacred mountain of Burkhan Khaldun. It crosses tundra, boreal forests, grasslands, and deserts in its journey to the Tatar Strait.

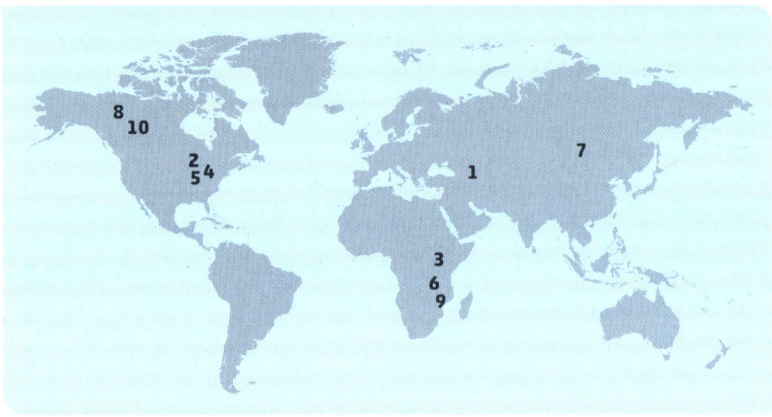

◀ **1. CASPIAN SEA, EUROPE/ASIA**
371,000 sq km (143,244 sq miles)

Located where Europe meets Asia, the Caspian is known as a sea due to its size and for political reasons. It's bordered by Russia, Azerbaijan, Iran, Turkmenistan, and Kazakhstan.

▲ **2. LAKE SUPERIOR, NORTH AMERICA**
82,100 sq km (31,699 sq miles)

There are five Great Lakes in North America, with Lake Superior being the largest. Around one-fifth of the world's fresh water is contained in the Great Lakes, with over half of this in Lake Superior.

TOP TEN
LARGEST
LAKES

Lakes are large bodies of water completely surrounded by land. Most lakes contain freshwater (water with little salt) and are continually replenished by rivers, streams, and precipitation, making up for water loss from outlets and evaporation. If a lake has no outlet and the climate is dry, evporation causes mineral levels to rise over time, making the water salty.

▲ **3. LAKE VICTORIA, EAST AFRICA**
68,870 sq km (26,591 sq miles)

Lake Victoria straddles Uganda, Tanzania, and Kenya, and its coastline is one of the most densely populated parts of East Africa. The lake is fed by the Kagera and Katonga Rivers, and its only outlet is the Victoria Nile on its northern coast.

◄ 7. LAKE BAIKAL, RUSSIA
31,500 sq km (12,200 sq miles)

Lake Baikal in eastern Siberia is the deepest and oldest lake on the planet at 20–25 million years old and stretching down to a gigantic depth of 1,620 m (5,315 ft). It's the world's largest freshwater lake by volume, containing one-fifth of Earth's unfrozen fresh water.

4. LAKE HURON, NORTH AMERICA ►
59,565 sq km (23,000 sq miles)

The second largest of the North American Great Lakes is Lake Huron. Despite being inland bodies of water, the Great Lakes can be treacherous to navigate, and it's estimated that Lake Huron is home to over 1,000 shipwrecks.

▲ 8. GREAT BEAR LAKE, CANADA
31,328 sq km (12,096 sq miles)

Located in the Fort Smith area of the Northwest Territories of Canada, Great Bear Lake straddles a transition zone where coniferous forests give way to tundra, with the northern waters crossing into the Arctic Circle.

◄ 5. LAKE MICHIGAN, NORTH AMERICA
57,753 sq km (22,300 sq miles)

Lake Michigan is even more dangerous than Lake Huron, with around a quarter of Great Lakes shipwrecks taking place there. It is the only Great Lake to lie completely within the US border.

◄ 6. LAKE TANGANYIKA, EAST AFRICA
32,900 sq km (12,700 sq miles)

Lake Tanganyika is the longest freshwater lake in the world, running over 644 km (400 miles) from Zambia at its southern end to Burundi on its northern shore.

▲ 9. LAKE MALAWI, EAST AFRICA
29,600 sq km (11,428 sq miles)

Lake Malawi makes up a large part of Malawi's border with Mozambique and Tanzania. The clear waters are home to hundreds of endemic fish species, which means they are only found in Lake Malawi.

▼ 10. GREAT SLAVE LAKE, CANADA
28,568 sq km (11,030 sq miles)

Great Slave Lake is located in the Northwest Territories of Canada. It is known for its large, rock-sloped bays and numerous islands, and the clear, frigid waters reach a depth of over 600 m (2,000 ft) in places.

TOP TEN
LONGEST CAVE SYSTEMS

Caves can meander for hundreds of miles underground, with features ranging from spectacular caverns to narrow passages. Cave systems are constantly being explored, and so could be longer than currently estimated.

▲ 2. OX BEL HA, QUINTANA ROO, MEXICO
436 km (270 miles)

Ox Bel Ha is also the longest underwater cave system. It's home to over 150 cenotes, which are sink holes or caverns that form when limestone bedrock is dissolved, causing a collapse of the land above.

▼ 3. SHUANGHEDONG CAVE SYSTEM, CHINA
401 km (249 miles)

The longest cave system in Asia, Shuanghedong has not yet been fully explored. It contains at least three underground rivers and numerous waterfalls.

1. MAMMOTH CAVE, KENTUCKY, USA ▶
685 km (426 miles)

Mammoth Cave in Kentucky is currently the longest cave system in the world thanks to the discovery of passages linking it to the Flint Ridge and Roppel Cave systems.

4. SISTEMA SAC ACTUN, QUINTANA ROO, MEXICO ▶
386 km (240 miles)

The Sac Actun is another vast underwater cave system dotted with cenotes. In 2011 divers discovered the bones of an ice-age mastodon along with a human skull that's thought to be around 12,000 years old.

▲ 5. JEWEL CAVE, SOUTH DAKOTA, USA
349 km (217 miles)

Jewel Cave gets its name from its numerous hanging crystals. Also visible are flowstones, which form when calcium carbonate-rich water flows over the walls in a sheet, leaving behind glossy-looking deposits of calcite.

6. OPTYMISTYCHNA, UKRAINE ▶
264 km (164 miles)

The "Optimistic Cave" is found near the village of Korolivka in western Ukraine. It is made up of a network of interconnecting passages, and so is known as a "maze cave".

8. CLEARWATER CAVE SYSTEM, MALAYSIA ▶
256 km (159 miles)

Found underneath the rainforests in the Sarawak region of Borneo, Clearwater (known locally as Gua Air Jernih) is the longest cave system in Southeast Asia, as well as being the largest in the world when measured by volume.

◀ 7. WIND CAVE, SOUTH DAKOTA, USA
260 km (162 miles)

This cave takes its name from the wind that blows in and out. Wind Cave's limestone caverns have a honeycomb pattern known as boxwork. This is where the rock has been eaten away by sulphuric acid, which forms when water reacts with the mineral gypsum. Due to the dry climate, there are few stalactites and stalagmites.

▲ 9. LECHUGUILLA, NEW MEXICO, USA
242 km (150 miles)

Speleothems are mineral deposits that form features such as stalactites. Lechuguilla has rare examples, including gypsum chandeliers and cave pearls.

▼ 10. FISHER RIDGE, KENTUCKY, USA
212 km (132 miles)

The Fisher Ridge cave system is described as having been rediscovered in 1981, as explorers have found evidence of human activity in the cave that dates back to around 3000 BCE.

TOP TEN
DEEPEST CANYONS

Canyons are formed over millions of years by the flow of rivers, the slow, gradual progression of erosion and weathering, and sometimes by tectonic uplift. They can eventually become huge trenches slicing into Earth's surface.

◄ **3. COLCA CANYON, PERU**
3,400 m (11,154 ft) deep

The Colca Canyon cuts through the Andes mountains in the south of Peru to a depth of around twice that of the Grand Canyon. It was formed by erosion from the Colca River, with the surroundings shaped by volcanic activity.

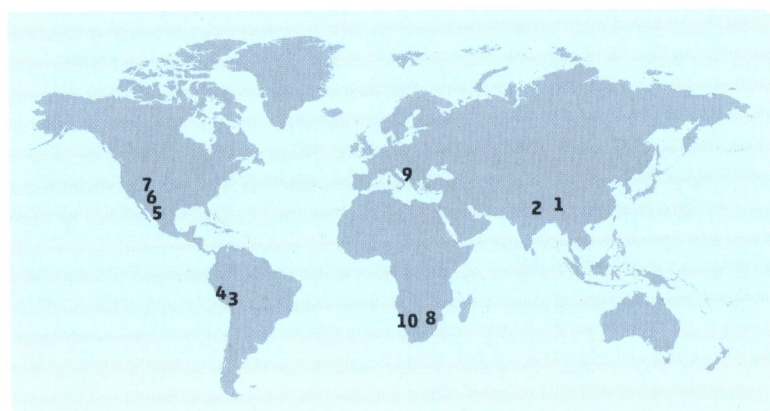

1. YARLUNG ZANGBO JIANG GRAND CANYON, EAST ASIA ▶
6,009 m (19,714 ft) deep

Not only is the Yarlung Zangbo the deepest canyon, it is also the world's longest canyon at around 500 km (300 miles).

◄ **2. KALI GANDAKI GORGE, NEPAL**
5,570 m (18,278 ft) deep

Situated in the Himalayas, the spectacular Kali Gandaki gorge (also known as the Andha Galchi) formed where a river cuts through an area of land that is rising rapidly due to tectonic movements (uplift).

◄ **4. COTAHUASI CANYON, PERU**
3,350 m (10,990 ft) deep

The Cotahuasi is the result of river erosion and glaciation. The area is remote and the river is dangerous, and there are few visitors apart from geologists and archaeologists conducting research.

◀ 5. COPPER CANYON, MEXICO
1,880 m (6,165 ft) deep

Copper Canyon is actually a group of six canyons that cut through the Sierra Madre Occidental, a mountain range situated in northwestern Mexico.

8. BLYDE RIVER CANYON, SOUTH AFRICA ▶
1,372 m (4,501 ft) deep

Whereas a lot of canyons are simply exposed rock, the Blyde River Canyon has abundant subtropical vegetation, making it the largest "green canyon".

◀ 6. GRAND CANYON, UNITED STATES
1,860 m (6,102 ft) deep

Probably the most famous canyon of all, the Grand Canyon was carved into the northwestern high plateau area of Arizona by the Colorado River. The walls expose rock layers that date back as far as 1.8 billion years, providing a detailed record of the slowly changing environment.

▼ 7. ANTELOPE CANYON, UNITED STATES
1,800 m (5,905 ft) deep

Slot canyons are very narrow gorges that are usually formed in soft rock. In the case of Antelope Canyon, situated on Navajo land in the state of Arizona, the sandstone was eroded by the rapid-flowing rush of repeated floods.

◀ 9. TARA RIVER CANYON, MONTENEGRO
1,310 m (4,300 ft) deep

The Tara River Canyon is the longest and deepest in Europe, stretching over 82 km (51 miles) from Montenegro to Bosnia and Herzegovina. The canyon is notable for containing over 80 large caves and featuring numerous sandy beaches under its high cliffs.

10. FISH RIVER CANYON, NAMIBIA ▶
550 m (1,805 ft) deep

Running for around 160 km (100 miles) through the south of Namibia, the Fish River Canyon is the longest canyon in Africa. It's made up of a wider upper canyon and a narrower lower canyon, with a river that flows only intermittently in the desert climate.

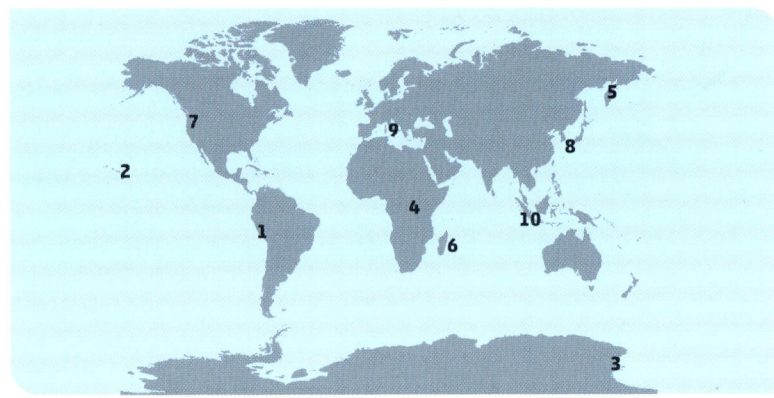

1. SABANCAYA, PERU ▶
5,960 m (19,554 ft)

High in the Peruvian Andes is Sabancaya, whose name translates as "tongue of fire". It's part of a three-volcano complex with Ampato and Hualca Hualca, neither of which are on record as being active. Between 29 May and 4 June 2023 Sabancaya averaged 24 explosions a day, sending gas and ash 2 km (1.2 miles) into the sky.

◀ 2. MAUNA LOA, HAWAII
4,170 m (13,681 ft)

Mauna Loa on Hawaii is a shield volcano. It is one of the most active volcanoes in the world, though not the most dangerous as it is less likely to erupt explosively. Instead, it erupts effusively, which means lava flows out from the volcano rather than being ejected violently as part of an explosive eruption.

▲ 3. MOUNT EREBUS, ANTARCTICA
3,794 m (12,448 ft)

Antarctica's Mount Erebus is the second highest volcano on the continent, after Mount Sidley at 4,200 m (13,780 ft), but is the most active, with regular emissions of gas and steam and occasional blasts that hurl lava bombs into the air.

VOLCANOES
AROUND THE WORLD

Volcanoes that are capable of erupting are described as active. Around 1,350 of the world's volcanoes are considered active, though many of these are currently dormant. If a volcano loses its connection to a magma supply, it will never erupt again and is described as extinct. Active volcanoes can be deadly, but not all eruptions are destructive – volcanoes in the sea can create new land, and volcanic ash can fertilize soil, helping plants grow.

4. NYIRAGONGO, DEMOCRATIC REPUBLIC OF CONGO ▶
3,470 m (11,385 ft)

Volcanoes made of layers of lava and ash are known as stratovolcanoes or composite volcanoes. Nyiragongo has a 2 km-wide crater that's normally filled with a lava lake. In 2002 the lake drained through openings in the mountain's flanks and almost totally destroyed the nearby city of Goma.

8. SAKURAJIMA, JAPAN ▶
1,117 m (3,665 ft)

Sakurajima is one of Japan's most active volcanoes. On average it erupts 100-200 times each year, and is capable of producing explosive eruptions that can send ash and gas plumes billowing into the sky to a height of 1.5 km (1 mile).

▲ 5. SHEVELUCH, RUSSIA
3,283 m (10,771 ft)

The largest and most active volcano on the Kamchatka Peninsula in the far east of Russia is Sheveluch. The volcano is notable for its lava domes, which form when thick lava reaches the surface but can't flow and so piles up around the vent.

6. PITON DE LA FOURNAISE, RÉUNION ▶
2,632 m (8,635 ft)

Situated on Réunion, an island in the Indian Ocean, Piton de la Fournaise is a shield volcano. Like Mauna Loa, it produces effusive eruptions, with basaltic lava flowing down its flanks at a speed of up to 8 m (26 ft) per second.

▲ 9. STROMBOLI, ITALY
924 m (3,032 ft)

Stromboli is a volcanic island located off the north coast of Sicily. It is one of four active volcanoes in Italy along with Etna, Vesuvius, and Vulcano. A Strombolian eruption is characterized by short explosive bursts where lava is projected tens or hundreds of metres into the air.

◀ 7. MOUNT ST HELENS, USA
2,549 m (8,363 ft)

Mount St Helens is a stratovolcano, and this type of volcano is more likely to produce explosive eruptions. On 18 May 1981, a build-up of pressure caused a massive explosion, destroying the volcanic cone and sending a huge pyroclastic flow into the valley. The air blast from the eruption blew down trees over a 500 sq km (200 sq mile) area.

10. ANAK KRAKATAU, INDONESIA ▶
155 m (509 ft)

One of the most famous eruptions ever was Krakatau in 1883, which created a crater beneath the sea in the Sunda Strait between the islands of Java and Sumatra. The smaller volcano, Anak Krakatau, formed in this caldera and began erupting in 1927, before emerging from the sea in 1929 and being active ever since.

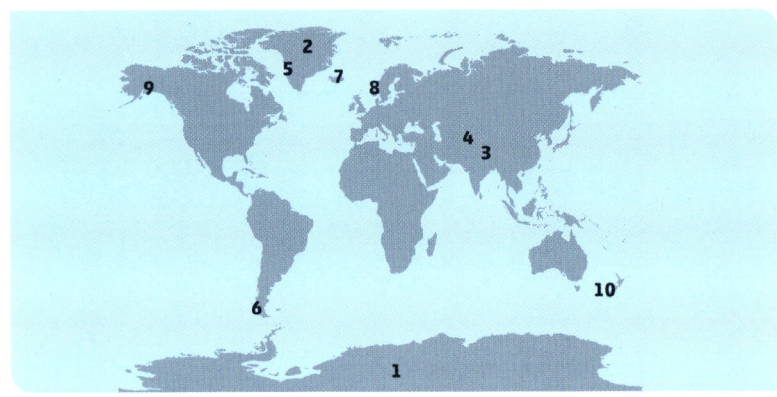

GLACIERS AROUND THE WORLD

Glaciers can be divided into broad groups based on size. Alpine glaciers are surrounded by mountain terrain and move incredibly slowly through the valleys below; ice sheets are the biggest in size and can cover entire mountains or continents. All types of glacier move slowly across the land, scouring the rock below and carving out valleys and other landforms.

1. ANTARCTIC ICE SHEETS, ANTARCTICA ▶

Almost all of Antarctica is covered by the West and East Antarctic ice sheets, which are divided by the Transantarctic Mountains. The East Antarctic Ice Sheet is the largest expanse of ice in the world. It is so heavy the land beneath it has sunk by around 2.5 km (1.6 miles).

◀ 2. GREENLAND ICE SHEET, GREENLAND

Around four-fifths of Greenland is covered by this ice sheet, making it the largest expanse of ice in the northern hemisphere. Loss of ice from the Greenland sheet through meltwater runoff and iceberg calving is one of the main causes of rising sea levels.

▲ 3. KHUMBU GLACIER, NEPAL

The highest glacier in the world is the Khumbu Glacier in the Himalayas. It is 15 km (9.3 miles) long and 4,900 m (16,100 ft) high at its end. Climbing Everest from Nepal takes you through the Khumbu, which is one of the most dangerous parts of the route.

4. FEDCHENKO GLACIER, TAJIKISTAN ▶

Outside of the polar regions, the largest alpine glacier at 77 km (48 miles) long is found in the Pamir Mountains in Tajikistan: the Fedchenko. The glacier is vital for the local population for drinking, agriculture, and energy.

8. JOSTEDAL GLACIER, NORWAY ▶

In mainland Europe the largest glacier is in the Jostedalsbreen National Park in Norway. The Jostedal Glacier is 60 km (37 miles) long and reaches an altitude of 1,957 m (6,420 ft) at its highest point. Norway's coastal fjords (long, narrow bodies of water) were carved out by glaciers like the Jostedal.

▲ 5. JAKOBSHAVN GLACIER, GREENLAND

The world's fastest glacier is the Jakobshavn in Greenland, which flows from Greenland's ice sheet into the ocean at speeds of up to 16 km (10 miles) a year. About 35 billion tonnes of icebergs break off the glacier's end every year.

6. SOUTHERN PATAGONIAN ICE FIELD, SOUTH AMERICA ▶

A great swathe of mountains in Chile and Argentina is covered by the Southern Patagonian Ice Field – the largest continuous mass of ice outside the polar regions. The Southern Patagonian Ice Field is about 13,000 sq km (5,000 sq miles) in area and feeds dozens of valley glaciers.

▲ 9. BERING GLACIER, ALASKA

This satellite image shows the largest glacier in North America – the Bering Glacier, which covers over 5,000 sq km (1,930 sq miles). Melting of the glacier may have caused more earthquakes to occur due to the reduction of downward pressure on faults along nearby plate boundaries.

◀ 7. VATNAJÖKULL GLACIER, ICELAND

Also called the Vatna glacier, Vatnajökull is a giant glacier (or "ice cap") that covers about 10 per cent of the country of Iceland. In places the ice is nearly 1 km (0.6 miles) thick. Under the glacier are several active volcanoes. Eruptions cause large floods called *jökulhlaups*, where meltwater builds up in a glacial lake and bursts through the sides.

10. TASMAN GLACIER, NEW ZEALAND ▶

Outside of Antarctica and South America, the only landmass in the southern hemisphere with glaciers is New Zealand. The largest is the Tasman Glacier (known as the Haupapa Glacier in Māori), which flows for 29 km (18 miles) on the east side of Mount Cook on the South Island.

IGNEOUS ROCKS

Igneous rocks form when magma (molten rock) cools and solidifies. There are two types of igneous rock: extrusive and intrusive.

EXTRUSIVE
These rocks form from lava flowing out of volcanoes and from rocks blasted out in explosive eruptions.

Rhyolite

Ignimbrite

Basalt

Porphyritic basalt

Amygdaloidal basalt

Banded rhyolite

Snowflake obsidian

Obsidian

Vesicular basalt

Ropy lava

Andesite

Pitchstone

Pumice

Lamprophyre

Porphyritic andesite

Trachyte

Agglomerate

Crystal tuff

INTRUSIVE
These rocks form when magma cools and solidifies underground.

Larvikite

Dunite

Kimberlite

Garnet peridotite

Troctolite

Pyroxenite

Pink granodiorite

Diorite

Syenite

Nepheline syenite

Layered gabbro

Leucogabbro

Olivine gabbro

Hornblende granite

Porphyritic granite

White microgranite

Pink microgranite

Anorthosite

Serpentinite

Porphyritic microgranite

Adamellite

Quartz porphyry

Feldspar pegmatite

Dolerite

Norite

Gabbro

White granodiorite

Mica pegmatite

Tourmaline pegmatite

Granophyre

Graphic granite

Bojite

METAMORPHIC
ROCKS

Metamorphic rocks are rocks that have had their original structure transformed by heat, pressure, or both. They can arise from any type of rock, including metamorphic rocks themselves.

FOLIATED
When the crystals that make up a metamorphic rock are arranged in layers, the rock is described as foliated.

Green slate

Black slate

Slate with pyrite

Garnet schist

NON-FOLIATED
When the crystals are arranged randomly rather than in layers, the metamorphic rock is described as non-foliated.

Blue marble

Fossiliferous slate

Phyllite

Muscovite schist

Biotite schist

Kyanite schist

Folded schist

Olivine marble

Green marble

Grey marble

Cordierite hornfels

Pyroxene hornfels

Garnet hornfels

Spotted slate

Chiastolite hornfels

Folded gneiss

Augen gneiss

Granular gneiss

Gneiss

Mylonite

Skarn

Migmatite

Amphibolite

Eclogite

Metaquartzite

Halleflinta

Granulite

SEDIMENTARY
ROCKS

Most sedimentary rocks are accumulations of sediment – pieces of rock, particles of sand, and other matter – that have settled at the bottom of a river, lake, or sea and been buried, compacted, and turned into rock. They also form when minerals crystallize out of water.

Breccia

DETRITAL

These sedimentary rocks form when mud, sand, and fragments of other rocks are compressed together.

Clay

Polygenetic conglomerate

Quartz conglomerate

Limestone breccia

Boulder clay

Loess

Siltstone

Mudstone

Calcareous mudstone

CHEMICAL

These rocks form from the crystallization of minerals that were dissolved in water.

Rock salt

Rock gypsum

Potash rock

Pisolitic limestone

Oolitic limestone

Dolomite

Tufa

Travertine

Banded ironstone

Oolitic ironstone

Internal structure

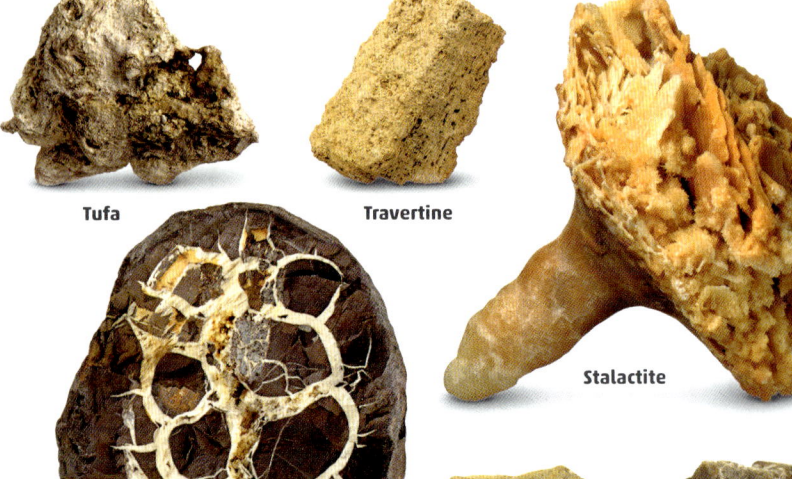

Stalactite

Outer surface

Septarian concretion

Chert

Flint

Pyrite nodule

Sandstone

Greensand

Red sandstone

Millet-seed sandstone

Limonitic sandstone

Micaceous sandstone

Pink orthoquartzite

Grey orthoquartzite

Greywacke

Arkose

Quartz gritstone

Red marl

Feldspathic gritstone

Red shale

Green marl

Black shale

Fossiliferous shale

ORGANIC
These sedimentary rocks are made largely from the remains of living things.

Nummulitic limestone

Shelly limestone

Crinoidal limestone

Coral limestone

Freshwater limestone

Red chalk

Chalk

Jet

Bryozoan limestone

Anthracite

Lignite

Bituminous coal

Sub-bituminous coal

Peat

Amber

MINERALS

Minerals are naturally occurring solid compounds with a specific structure and chemical formula. Scientists split minerals into groups based on their chemical composition.

SILICATES

This large group is made up of minerals containing the elements silicon and oxygen combined with other elements.

Olivine

Grossular garnet

Topaz

Chondrodite

Titanite

Euclase

Beryl

Eudialyte

Hemimorphite

Epidote

Tourmaline

Clinozoisite

Vesuvianite

Ilvaite

Axinite

Benitoite

Aegirine

Spodumene

Diopside

Jadeite

Hornblende

Augite

Rhodonite

Riebeckite

Tremolite

Neptunite

Talc

Chrysocolla

Phenakite

Lepidolite

Clinochlore

Prehnite

Fluorapophyllite

Pyrophyllite

Microcline

Orthoclase

Lazurite

Sodalite

Heulandite

Labradorite

Scolecite

Stilbite

Opal

Quartz

Chalcedony

OXIDES AND HYDROXIDES
Minerals in this group contain oxygen or hydroxide combined with other elements.

Zincite

Cuprite

Perovskite

Uraninite

Franklinite

Haematite

Chrysoberyl

Samarskite

Limonite

Diaspore

NATIVE ELEMENTS
These minerals are made purely of one element. They include metals, semimetals, and nonmetals.

Graphite (carbon)

Diamond (carbon)

Arsenic

Gold

Bismuth

SULPHIDES AND SULPHOSALTS

These minerals contain the nonmetal element sulphur combined with metals or semimetals.

Galena

Cinnabar

Stibnite

Acanthite

Covellite

Realgar

Molybdenite

Arsenopyrite

Pyrite

Chalcopyrite

Millerite

Bournonite

Proustite

Enargite

CARBONATES

This group contains compounds in which one or more metal or semimetal elements combine with the carbonate ion.

Aragonite

Rhodochrosite

Siderite

Calcite

Artinite

Azurite

Aurichalcite

SULPHATES

The minerals in this group are made of one or more metals combined with the sulphate ion.

Celestine

Barite

Anglesite

Brochantite

Chalcanthite

Linarite

PHOSPHATES

The minerals in this group contain metals combined with the phosphate ion.

Lazulite

Pyromorphite

Autunite

Vivianite

Turquoise

Variscite

Wavellite

HALIDES

In this group of minerals, metals combine with halogens – elements such as fluorine and chlorine.

Carnallite

Sylvite

Atacamite

Fluorite

Diaboleite

OTHER GROUPS

Other, less common groups of minerals include vanadates, molybdates, borates, chromates, tungstates, and arsenates.

Wulfenite

Descloizite

Vanadinite

Nitratine

Colemanite

Clinoclase

Crocoite

Wolframite

Adamite

Mimetite

Bayldonite

GEMSTONES

Gemstones are rocks and minerals that can be polished or cut to make jewellery. Some gemstones come in a range of colours, which may be due to impurities.

Goshenite (beryl)

Colourless sapphire

Rock crystal (quartz)

Albite

Milky quartz

Haematite

Colourless orthoclase

Fluorite

Petalite

Euclase

Phenakite

Datolite

Taaffeite

Celestine

Pearls

Hambergite

Beryllonite

Pale pink diamond

Elephant ivory

Morganite (beryl)

Rose quartz

Unpolished tugtupite

Unpolished pink grossular (garnet)

Spinel

Fluorite

Padparadscha (corundum)

Red beryl

Watermelon tourmaline

Rubellite (tourmaline)

Ruby (corundum)

Unpolished red-pink diamond

Almandine (garnet)

Pyrope (garnet)

Hessonite (grossular garnet)

Carnelian (chalcedony)

Red coral

Vegetable ivory

Banded rhodochrosite

Sardonyx (chalcedony)

Amber

Aventurine quartz

Rhodonite

Baryte

Axinite

Fire agate (chalcedony)

Hypersthene

Spessartine (garnet)

Unpolished jasper (chalcedony)

Smoky quartz

Aragonite

Unpolished diamond

Unpolished pyrite

Alexandrite (chrysoberyl)

Brown diamond

Dravite (tourmaline)

Heliodor (beryl)

Titanite

Tiger's eye (chatoyant quartz)

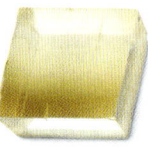
Unpolished meerschaum

Gypsum

Unpolished quartz with inclusions

Anglesite

Sinhalite

Cerussite

Danburite

Prehnite

Unpolished sunstone (oligoclase)

Cassiterite

Calcite

Sphalerite

Yellow orthoclase

Spinel

Pink sapphire (corundum)

Taaffeite

Amethyst (quartz)

Zoisite

Jet

Kunzite (spodumene)

Dumortierite quartz

Bort diamond

Obsidian

Black coral

Unpolished staurolite

Spinel

Iolite

Sapphire (corundum)

Hauyne

Moonstone (orthoclase)

Topaz

Apatite

Blue John (fluorite)

Polished sodalite

Opal

Labradorite

Kyanite

Zircon

Stained howlite

Lazulite

Unpolished uvarovite (garnet)

Aquamarine (beryl)

Translucent sodalite

Lapis lazuli

Ruby (corundum)

Benitoite

Diopside

Malachite

Jadeite (jade)

Nephrite (jade)

Indicolite (tourmaline)

Agate (chalcedony)

Turquoise

Azurite-malachite

Mother-of-pearl

Dark green diamond

Unpolished chrysoprase (chalcedony)

Emerald (beryl)

Smithsonite

Grandidierite

Phosphophyllite

Amazonite (microcline)

Green sapphire (corundum)

Chrysocolla

Andalusite

Moldavite (tektite)

Kornerupine

Sillimanite

Brazilianite

Unpolished pale green diamond

Dioptase

Peridot (olivine)

Tsavorite (garnet)

Schorl (tourmaline)

Amblygonite

Dolomite

Achroite (tourmaline)

Unpolished diamond

Enstatite

Serpentine

Demantoid (andradite garnet)

Unpolished bloodstone (chalcedony)

Yellow sapphire (corundum)

Citrine (quartz)

Green and yellow tourmaline

Fluorite

Vesuvianite

Scheelite

Demantoid (andradite garnet)

Epidot

Sphalerite

FOSSILS

Only a small fraction of extinct species have left evidence of their existence as fossils. Even so, fossils have helped scientists work out the tree of life on Earth. Fossils also help us find out the age of rocks.

Turtle

Plesiosaur flipper

VERTEBRATES

Vertebrates are animals with a backbone and an internal skeleton. Their bones and other hard body parts make well-preserved fossils.

Crow shark tooth

Ivory tusks

Limestone rock

Fish (*Furo*)

Fish (*Diplomystus*)

Fish (*Hoploteryx*)

Mammoth

INVERTEBRATES

Invertebrates are mostly small animals without internal skeletons, such as snails, crabs, and insects. Many invertebrate fossils are shells of sea creatures.

Sponge **Horn coral** **Trilobite**

Decapod

Crab

Horseshoe crab

Whip scorpion

Ammonites

Mussel

Bivalve (*Neocrassina*)

Bivalve (*Teredina*)

Bivalve (*Wilkingia*)

Sea snail

Tulip snail

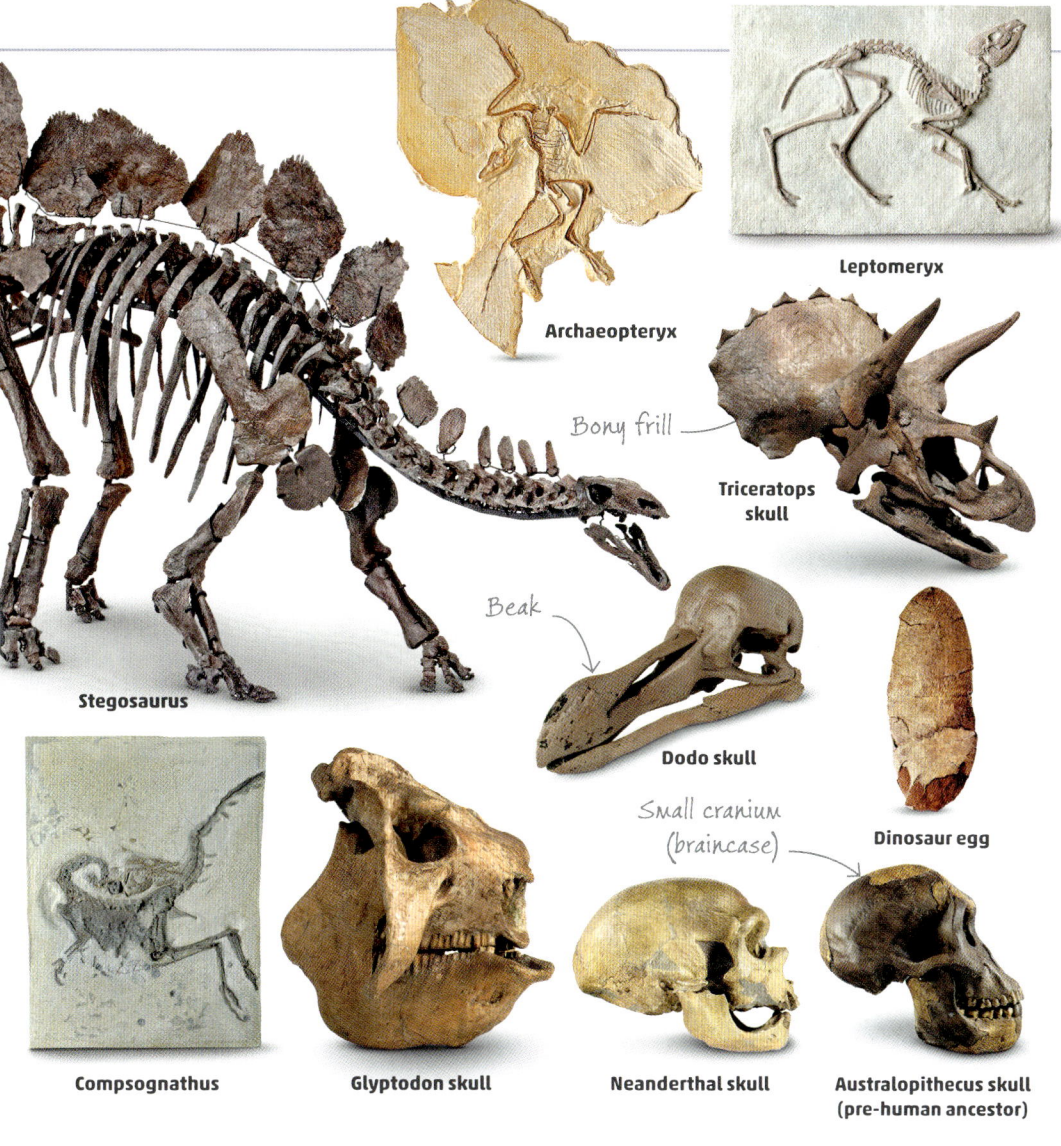

Archaeopteryx

Leptomeryx

Triceratops skull

Bony frill

Beak

Stegosaurus

Dodo skull

Dinosaur egg

Small cranium (braincase)

Compsognathus

Glyptodon skull

Neanderthal skull

Australopithecus skull (pre-human ancestor)

PLANTS
Plant fossils include seeds, imprints of leaves, and tree trunks that have turned to solid rock. Coal, which is used as a fuel, consists entirely of fossilized plant material.

Tree fern

Horsetail

Seed fern

Trigonocarpus seeds

Sequoia tree

Monkey puzzle tree

Tree trunk

Amber

Plane tree

Oak

Palm fruit

Crinoid

Ammonite

Belemnite

Shell

Sea urchin

Brittle star

Sea star

Blastoid

Cystoid

Stylophoran

GREENLAND

NORTH
AMERICA

ATLANTIC
OCEAN

AFRICA

PACIFIC OCEAN

SOUTH
AMERICA

THE CONTINENTS

Earth's land is divided into major areas called
continents. By convention, there are seven named
continents, though Europe and Asia are in fact a single
landmass (Eurasia). Earth also has at least 900,000
islands, the largest of which is Greenland. The exact
number of islands is unknown as many are so small. The
country with the most is Sweden, with 267,570 islands.
Fewer than 1 per cent of these are inhabited.

GLOSSARY

AGRICULTURE
The practice of farming – growing crops and raising livestock.

ALGAE
Simple, plantlike organisms that live in water and make their food by photosynthesis.

ALTITUDE
The vertical distance between an object and the Earth's surface or sea level.

AMMONITE
The fossilized coiled shell of an extinct cephalopod mollusc (a relative of squid).

AMPHIBIAN
A cold-blooded, backboned animal that lives part of its life in water and part on land, such as a frog.

ANTICLINE
An arch-like, upward fold of originally flat strata formed by horizontal compression.

AQUATIC
A term used to describe animals and plants that live in water.

ASTEROID
A relatively small, irregular rocky object orbiting the Sun.

ASTHENOSPHERE
The soft layer of upper mantle on which the tectonic plates move.

ATMOSPHERE
The layer of gas that surrounds a planet, retained by gravity.

ATOLL
A ring-shaped coral island or a ring-shaped group of coral islands surrounding a lagoon.

ATOM
The smallest particle of an element.

AURORA
A pattern of light in the night sky in Arctic and Antarctic regions, caused by the interaction between electrically charged particles from the Sun and Earth's magnetic field.

BACTERIA
Microscopic single-celled organisms that make up one of the main kingdoms of life on Earth. Many bacteria are helpful but some cause diseases.

BASALT
Earth's most common volcanic rock, which usually originates as solidified lava. Basalt is glassy to fine-grained (composed of very small crystals).

BEDROCK
The layer of solid rock below the soil.

BIOME
A major division of the living world, such as rainforest, desert, or temperate grassland. Each biome has its own distinctive climate, vegetation, and animal life.

BIOMINERALIZATION
The process by which living creatures produce minerals.

BIOSPHERE
The parts of Earth where life exists.

BIVALVE
An aquatic mollusc, such as an oyster or clam, with two hinged shells that cover its whole body.

BLACK SMOKER
A hot spring containing dark minerals that erupts from the ocean floor.

BRECCIA
A sedimentary rock made up of angular fragments, called clasts, cemented by minerals.

CALDERA
A giant crater formed when a volcano collapses into its emptied magma chamber after an eruption.

CALVING
The process by which a glacier creates icebergs by shedding ice blocks into a sea or lake.

CAMOUFLAGED
Coloured, patterned, or shaped to match the surroundings. Animals use camouflage to hide.

CANYON
A rocky valley with steep sides.

CARBON DIOXIDE
A gas found in air. Animals breathe out carbon dioxide as a waste product, but plants absorb it.

CELL
A tiny unit of living matter. Cells are the building blocks of all living things.

CINDER CONE
A relatively low-lying volcano formed from cinders and ash that have fallen from an eruption cloud.

CLIMATE
The most common weather conditions in an area over a period of time.

CLIMATE CHANGE
Long-term changes to weather patterns and average temperatures on a global or regional scale.

COMET
A space object made of ice and dust that travels from the outer solar system and passes around the Sun.

COMPOUND
A chemical substance in which two or more elements have bonded together.

CONDENSATION
When a gas turns into a liquid.

CONGLOMERATE
A sedimentary rock made up of rounded, gravel-size particles cemented together by minerals.

CONIFER
A cone-bearing tree. Nearly all conifers, such as pines and firs, have needle-like leaves and are evergreen.

CONTINENT
One of the seven large landmasses on Earth, mainly surrounded by sea.

CONTINENTAL CRUST
The part of Earth's crust that forms the continents. It is less dense and thicker than oceanic crust.

CONTINENTAL SHELF
The submerged edge of a continent that lies beneath shallow coastal seas.

CONVECTION
The circulation of gases, liquids, or heat-softened rock, driven by temperature differences.

CONVERGENT BOUNDARY
Where two tectonic plates are moving together.

CORALS
Marine animals that may have a hard base made of limestone. Many live in large groups or colonies that build coral reefs.

CORE
Earth's innermost layer. It consists of a liquid outer core and a solid inner core, both made of nickel and iron.

CORIOLIS EFFECT
The tendency of winds and ocean currents to bend towards the east or west due to Earth's rotation.

CRATER
A bowl-shaped depression around the vent of a volcano or a circular depression caused by a meteorite impact.

CRUST
The cool, rocky outer shell of Earth.

CRYSTAL
A solid with an orderly internal atomic structure that sometimes shows geometric forms with flat faces.

CUMULONIMBUS
A very tall cloud that produces heavy rain, lightning, and hail.

CYCLONE
A pressure system in which air circulates around an area of low pressure. Tropical cyclones are also known as hurricanes.

DECIDUOUS
A word used for trees or shrubs that shed their leaves every year, usually in autumn.

DELTA
An area of land formed from sediment deposited where a river meets the sea.

DENDRITIC
A term applied to minerals occurring naturally in a branching, treelike form within rocks.

DENSITY
The mass of an object divided by its volume. Dense objects are heavy.

DEPOSITION
The laying down of material such as sand and gravel in new locations, usually by wind, water, or ice.

DESERT PAVEMENT
A rocky or stony surface layer found in many deserts.

DIVERGENT BOUNDARY
Where two tectonic plates are moving apart.

DNA
Deoxyribonucleic acid, a chemical substance that stores genetic information and is found inside cells.

DORMANT
A volcano is described as dormant if it has the potential to erupt but is not currently erupting.

DOWNWELLING
The sinking down of water from the surface of the ocean.

DROUGHT
A long period of unusually low rainfall resulting in a lack of water and very dry conditions.

DYKE
A sheetlike intrusion of igneous rock that cuts across existing rock structures.

DYNAMO
A device that changes movement or mechanical energy into electrical energy.

EARTHQUAKE
A violent shaking of Earth's surface caused by sudden, massive movements of areas of crust.

ECOSYSTEM
A community of living things and their environment. An ecosystem may be as small as a pond or as large as a rainforest.

ELEMENT
A simple substance made of atoms that are all of the same kind.

ELEVATION
The height of a place or object above land or sea level.

EMISSION
Something released into the air, such as gas, heat, or light.

EPICENTRE
The point on Earth's surface directly above the point of origin of an earthquake.

EQUATOR
An imaginary circle around the centre of Earth, dividing it into the northern and southern hemispheres (halves).

EROSION
The wearing away of rock by natural forces, such as flowing water, wind, or glaciers.

ESTUARY
A funnel-shaped stretch of mud and tidal water where a river meets the sea.

EVAPORATION
A process by which a liquid changes into a gas.

EVERGREEN
A term for plants that bear leaves throughout the year.

EVOLUTION
The gradual process of change in living things over many generations.

EXOSKELETON
A hard outer skeleton, such as that of an insect.

EXTINCTION
The disappearance from Earth of the last living representative of a species.

EXTRUSIVE ROCK
Rock formed from lava that flowed onto Earth's surface or was ejected in an ash cloud.

FAULT
A fracture in Earth's crust, where one slab of rock slides past another.

FELDSPAR
A common type of silicate mineral found in igneous rock.

FERTILIZER
A substance that helps plants grow better when added to soil.

FISSURE
A deep, narrow gap, such as between rocks.

FJORD
A former glacial valley on the coast that has become an inlet of the sea.

FLASH FLOOD
A sudden flood caused by heavy rain.

FLOODPLAIN
A flat plain next to a river that is liable to become flooded.

FOLD
A geological structure in which originally flat layers of rock have bent under compression.

FOLIATION
The formation of leaflike layers in metamorphic rock.

FOSSIL
The preserved remains or traces of animals or plants from an earlier time.

FOSSIL FUEL
Fuel derived from fossils, such as coal or petroleum.

FULGURITE
Naturally formed tubes of glassy minerals formed when lightning strikes the ground.

FUNGUS
A living thing that absorbs food from living or dead matter around it. Mushrooms and toadstools are the fruiting parts of fungi.

GALAXY
A rotating cloud of stars, gas, and dust held together by gravity.

GEMSTONE
A jewel or stone used in jewellery that is prized for its durability, beauty, and rarity.

GEODE
A rock containing a hollow cavity lined with crystals.

GEOTHERMAL
A term for heat from below ground.

GEYSER
An intermittent jet of hot water and steam that erupts from volcanically heated rock.

GLACIER
A mass of ice formed from compacted snow that flows slowly downhill.

GONDWANA
An ancient supercontinent that included land now in South America, Africa, Antarctica, Australia, and India.

GORGE
A narrow, deep valley, usually with vertical cliffs on either side.

GRANITE
One of the main igneous rocks found in continental crust.

GRAVITY
The force that attracts one object to another and prevents things floating off into space.

GREENHOUSE GAS
A gas, such as carbon dioxide, that traps heat in Earth's atmosphere.

GROUNDWATER
Water present beneath Earth's surface in the rocks and soil.

GYRE
A system of rotating ocean currents.

HABIT
The distinctive external shape of a crystal or group of crystals.

HABITAT
The area where an organism naturally makes its home.

HANGING VALLEY
A valley, usually carved by a glacier, that enters high up the side of a deeper valley.

HARDNESS
Of a mineral, the extent to which it can resist scratching or abrasion.

HEADWATER
The upper portion of any river or stream, close to its source.

HERBIVORE
An animal that eats plants.

HIBERNATION
A resting state like very deep sleep that occurs in some animals in winter.

HOTSPOT
An area beneath Earth's crust where the mantle is particularly hot, creating a zone of volcanic activity.

HUMUS
A dark-coloured substance found in soils, derived from dead plants, microorganisms, and animals.

HURRICANE
A huge, rotating, tropical storm in which wind speeds exceed 119 kph (74 mph). It is also called a tropical cyclone and (especially in eastern Asia) a typhoon.

ICE AGE
A cold period in Earth's history when ice sheets and glaciers get larger.

ICE FLOE
A large area of floating sea-ice.

ICE SHEET
A very large, flowing mass of ice that permanently covers a land area, as in Antarctica or Greenland.

ICE SHELF
A floating extension of an ice sheet or glacier over the ocean.

ICEBERG
Part of a glacier or ice shelf that has broken off and floated out to sea.

IGNEOUS INTRUSION
A body of igneous rock that formed when magma cooled and solidified underground.

IGNEOUS ROCK
Rock formed by the cooling of molten magma or volcanic lava.

INSECTIVORE
An animal that mainly eats insects or other invertebrates.

INTRUSIVE ROCK
Igneous rock that has solidified below the surface, cooling slowly enough to allow larger crystals to form. A body of intrusive igneous rock is called an igneous intrusion.

INVERTEBRATE
An animal without a backbone, such as an insect or worm.

IRRIGATION
Supplying land with water so that crops and plants will grow.

ISOSTATIC REBOUND
The rise of a land mass after the huge weight of an ice sheet is removed.

JET STREAM
A narrow band of strong wind high in the atmosphere.

LATITUDE
A measure of how far north or south from the Equator a location is. The Equator has a latitude of 0° and the North Pole is at 90°.

LAURASIA
An ancient supercontinent that included land now in North America, Europe, and Asia.

LAVA
Molten rock that erupts from a volcano.

LAVA BOMB
A blob of lava hurled through the air during a volcanic eruption.

LIMESTONE
Sedimentary rock made of calcium carbonate, formed from the shells of sea animals or from chemical deposition.

LITHOSPHERE
Earth's solid outer layer, made of the crust and the uppermost part of the mantle. The lithosphere is divided into tectonic plates.

LUSTRE
The way a mineral reflects light and the extent of its sheen.

MAGMA
Molten rock that lies within or beneath Earth's crust.

MAGMA CHAMBER
A reservoir of magma within or below a volcano.

MAGNETIC FIELD
The area around a magnet or Earth where magnetic force is felt.

MAGNETOSPHERE
The region around Earth (or any other planet) dominated by the planet's magnetic field.

MAMMAL
A warm-blooded vertebrate (animal with a backbone) that feeds its young on milk and usually has a covering of fur.

MANGROVE
A type of tree that grows on tropical coasts and has stilt-like roots to stay above the water.

MANTLE
The rocky layer of Earth between the crust and the core. It makes up 84 per cent of Earth's volume.

MANTLE PLUME
A hotter region of the mantle that flows slowly upwards to form a hotspot beneath the crust.

MARSUPIAL
A mammal that gives birth to very undeveloped young and that usually carries babies in a pouch.

MEANDER
A bend in a river created by deposition on the slow side and erosion on the faster side.

MESOSPHERE
The layer of the Earth's atmosphere between the stratosphere and thermosphere, at an altitude of about 50–80 km (30–50 miles).

METAMORPHIC AUREOLE
A zone of rock around a body of magma that has been altered by contact metamorphism.

METAMORPHIC ROCK
Rock transformed by heat, intense pressure, or both.

METEOR
A small rock from space that vaporizes as it falls through Earth's atmosphere, creating a shooting star.

METEORITE
A rock from space that has fallen to Earth's surface without completely burning up.

METHANE
A natural gas that burns easily and is used as fuel. It is a greenhouse gas.

MICROORGANISM
An organism too small to be seen with the naked eye, such as a bacterium.

MID-OCEAN RIDGE
A submarine mountain chain created by volcanoes erupting from a rift in the ocean floor.

MIGRATION
A long journey undertaken by an animal to reach a new habitat. Many birds migrate every year between their summer and winter homes.

MILKY WAY
The galaxy of stars that contains the Sun and our solar system.

MINERAL
A natural, inorganic, solid substance that has a characteristic crystal structure. Rocks are made of minerals.

MOHS SCALE
A measure of the hardness (resistance to scratching) of minerals, ranging from 1 to 10.

MOLECULE
A group of atoms bonded together.

MOLLUSC
A soft-bodied invertebrate that may have a hard shell. Snails, clams, and octopuses are molluscs.

MOLTEN
Melted (in liquid form).

MOON
A small rocky world that orbits a planet.

MORAINE
Rock debris carried by a glacier or piled up at its sides or end.

MUD POT
A hot spring created by acidic water that turns rock to mud.

NATIVE ELEMENT
A chemical element that occurs in a pure state in nature.

NEAP TIDE
A tide in which the difference between high and low tide is at its smallest.

NEBULA
A cloud of gas and/or dust in space.

NECTAR
A sugary liquid produced by flowers. Bees collect nectar in order to make honey.

NITROGEN
The gas that forms 78 per cent of Earth's atmosphere.

NOCTURNAL
Active at night but inactive during the day.

NODULAR
Containing rounded lumps of minerals or other materials. For example, flint in chalk.

NUCLEAR ENERGY
Energy released by the decay (breakdown) of radioactive elements or by the fusion of atomic nuclei.

NUTRIENTS
Substances that animals and plants take in and that are essential for life and growth.

OCEANIC CRUST
The part of Earth's crust that underlies most of the world's oceans. It is thinner and denser than continental crust.

ORBIT
To circle a planet or star. For example, Earth orbits the Sun, and the Moon orbits Earth.

ORE
A rock from which metal can be mined.

ORGANISM
A living thing.

OUTWASH
The deposit of sand, gravel, and other materials carried by meltwater from a glacier.

OXBOW LAKE
A river meander cut off from the main flow, forming a U-shaped lake.

OXIDE
A chemical containing oxygen and one other element, such as a metal.

OXYGEN
A gas that makes up 21 per cent of Earth's atmosphere. Most living things take in oxygen from the air and use it to release energy from food in a process called respiration.

PANGAEA
An ancient supercontinent that included almost all the present-day continents.

PERMAFROST
Permanently frozen ground below the surface of the soil.

PERMEABLE
A term for a substance that allows liquids or gases through it.

PHOTOSYNTHESIS
The process by which plants use the Sun's energy to make food molecules from water and carbon dioxide.

PHYTOPLANKTON
The tiny, mostly single-celled organisms that live in the surface waters of oceans and lakes and are the base of most aquatic food chains.

PILLOW LAVA
Pillow-shaped mounds of rock formed from lava that has been ejected underwater.

PLANETESIMAL
One of millions of rocky objects of variable size believed to have been present in the early solar system and that later clumped together to create the planets.

PLANKTON
Organisms that float in water and drift with the currents. Most plankton are small.

PLATEAU
A level area of land at high altitude.

POLLUTANT
A substance that contaminates air, water, or soil.

PRECIPITATION
Water that reaches Earth's surface from the atmosphere, including rain, snow, hail, and dew.

PREDATOR
An animal that hunts other animals for food.

PREHISTORIC
Existing before written records.

PREY
An animal hunted by other animals for food.

PRISMATIC
A term to describe crystals in which the parallel rectangular faces form prisms.

PYROCLASTIC
Consisting of or containing volcanic rock fragments. Pyroclastic flows are fast-moving, sometimes deadly clouds of hot gases and debris.

PYROCLASTIC FLOW
An avalanche of very hot rock and dust that cascades down the flank of an erupting volcano.

RAIN SHADOW
An area of low rainfall downwind of a mountain range, caused by air losing moisture as it rises over the windward side of the range.

REFRACTION
The bending of light rays when they pass from one medium, such as air, to another, such as water or glass.

REPRODUCTION
Production of new offspring.

REPTILE
A cold-blooded, scaly-skinned animal such as a snake or lizard.

RIFT
A crack in Earth's crust where two tectonic plates are pulling apart.

RIFT VALLEY
A valley formed by collapse of parts of Earth's crust at the boundary between diverging tectonic plates.

SANDSTONE
A rock made of sand grains cemented together by other minerals.

SEA STACK
A tall column of rock rising out of the sea near a coastline, a resistant remnant of surrounding cliffs that have otherwise disappeared due to erosion.

SEAMOUNT
An underwater volcano that is not high enough to breach the surface and form an island.

SEDIMENT
Small fragments of rock, sand, or mud that settle in layers, usually underwater.

SEDIMENTARY ROCK
Rock formed from compressed and hardened sediments.

SEISMIC
Relating to earthquakes. A seismic wave is a shockwave generated by an earthquake.

SEISMOMETER
An instrument for recording seismic waves from earthquakes. Also called a seismograph.

SHIELD VOLCANO
A very wide volcano with gently sloping sides built up from layers of fast-flowing lava.

SILICATE
A rock or mineral made of silicon and oxygen atoms combined with metal atoms. Silicates make up most of Earth's crust and mantle.

SILL
A roughly horizontal, sheet-like igneous intrusion that usually forms when igneous rock forces its way between layers of existing sedimentary rocks.

SILT
Fine particles carried by water.

SOLAR SYSTEM
The system of planets, moons, comets, and asteroids orbiting the Sun.

SOLAR WIND
A stream of charged particles released from the upper atmosphere of the Sun.

SPECIES
A type of organism, such as a cheetah or a giraffe. Members of the same species can breed together in the wild.

SPRING TIDE
A tide that occurs when the effects of the Moon and Sun reinforce each other, producing the highest high tide and the lowest low tide.

STALACTITE
A deposit of calcium carbonate hanging from the roof of a cave.

STALAGMITE
A deposit of calcium carbonate on the floor of a cave.

STAR
A massive sphere of glowing plasma (highly energized gas) that generates energy by nuclear fusion in its core.

STORM SURGE
An abnormal rise in sea level caused by a storm.

STRATOVOLCANO
A conical volcano built up from layers of volcanic ash and slow-flowing lava.

STRIATIONS
Grooves and scratches that are left in the bedrock as a glacier moves over them.

SUBDUCTION
When one tectonic plate sinks beneath another.

SUBSOIL
The layer of soil immediately beneath the topsoil.

SULPHUR
A yellow element, common in rocks and often erupted from volcanoes.

SUPERCELL
A huge storm cloud with a rotating core that may generate a tornado.

SUPERCOOLED
Describes a liquid that is cooled to below its normal freezing point without becoming solid.

SUPERHEATED
Describes a liquid heated under pressure to above its normal boiling point without turning to gas.

SUPERVOLCANO
A gigantic volcano.

SURFACE TENSION
A force that creates a slightly elastic skin on the surface of water. Surface tension pulls water into droplets.

SYMBIOSIS
A close, usually beneficial relationship between two species that live together.

TECTONIC PLATE
A gigantic fragment of Earth's crust. Tectonic plates move slowly over millions of years.

TEMPERATE
A term for a climate that is neither very hot nor very cold.

THRUST FAULT
A major fracture in Earth's crust where one section is pushed above another.

TIDAL BORE
A single large wave that forms when a rising tide enters a narrowing channel such as an estuary.

TOPSOIL
The uppermost layer of soil, containing minerals and organic matter in which plants grow.

TRADE WINDS
Winds on either side of the equator that blow west.

TRANSFORM BOUNDARY
A boundary between tectonic plates where the plates slide horizontally past each other.

TRANSFORM FAULT
A boundary between two slabs of Earth's crust where they slide sideways relative to each other.

TRILOBITE
An extinct marine animal with a segmented body, an exoskeleton, and many pairs of jointed legs. Trilobites went extinct around 252 million years ago.

TROPICAL CYCLONE
Another name for a hurricane.

TROPICS
The regions of Earth close to the Equator, where the climate is warm all year round.

TROPOSPHERE
The lowest, densest layer of the atmosphere, where clouds form and most weather phenomena occur.

TSUNAMI
A fast-moving, destructive wave generated most often by an earthquake on the sea floor.

TYPHOON
A hurricane or tropical cyclone in the western Pacific or Indian Oceans.

VERTEBRATE
An animal with a backbone.

VISCOSITY
Resistance to flow in fluids. The higher the viscosity of a fluid, the more sluggishly it flows.

WATER TABLE
The upper surface of groundwater. A rise in the water table after torrential rain can cause flooding.

WATER VAPOUR
An invisible gas formed when water evaporates into the air.

WEATHERING
The breaking down of rocks and minerals by rain, sunlight, ice, and other climatic effects.

INDEX

ACKNOWLEDGMENTS

Dorling Kindersley would like to thank the following for invaluable help and expert knowledge. Professor Jürgen Adam, Professor in Structural Geology, and Kevin D'Souza at Royal Holloway University of London for setting up and running sandbox simulations of mountain formation. Joel Rees-Jones, Project Manager at LIFEDeeRiver in Wales, for setting up river and delta simulations in an Emriver table. Dr Daniel Whiter, lecturer in Space Environment Physics at the University of Southampton, for setting up an aurora simulation. Simon Cohen, Jacqueline Cohen, Duncan Barrington, Claire Alway, and Sam Cohen at Fossilsandgemstones.co.uk for providing rocks, minerals, and fossils. Professor John Brodholt and Dr Andrew R. Thomson at University College London for showing us UCL's rock and mineral collection.

For Smithsonian Enterprises
Avery Naughton, Licensing Coordinator; Paige Towler, Editorial Lead; Jill Corcoran, Senior Director, Licensed Publishing; Brigid Corcoran, Vice President of New Business and Licensing; Carol LeBlanc, President.
Matthew T. Miller, Museum Specialist, for cconsultancy.

Thanks also to Tom Morse for image retouching; Pushpak Tyagi for DTP design; Usman Ansari for photoshop assistance; and Steve Setford for editorial support.

Picture Credits
The publisher would like to thank the following for their kind permission to reproduce their photographs (a-above; b-below; c-centre; f-far; l-left; r-right; t-top):

1 Dorling Kindersley: Dreamstime.com: Mario Lopes / Malopes. **Getty Images / iStock:** ChrisGorgio (bolts). **Science Photo Library:** Martin Rietze (c). **2-3 Dorling Kindersley:** Dreamstime.com: Mario Lopes / Malopes (background). **Getty Images / iStock:** ChrisGorgio (bolts). **3 Alamy Stock Photo:** heyengel (c). **4-5 Dorling Kindersley:** Dreamstime.com: Mario Lopes / Malopes. **5 NASA. Shutterstock.com:** Wirestock Creators (tr). **6-7 Dorling Kindersley:** Dreamstime.com: Mario Lopes / Malopes (background). **6 Alamy Stock Photo:** Mario Deambrogio (tc). **Kenneth G. Libbrecht. Shutterstock.com:** Blue Planet Studio. **7 Getty Images:** imageBROKER / Peter Giovannini. **8-9 Dorling Kindersley:** Dreamstime.com: Mario Lopes / Malopes (background). **8 Getty Images / iStock:** ChrisGorgio. **9 Getty Images / iStock:** ChrisGorgio (screws). **NASA. 10-11 Science Photo Library:** Natural History Museum, London. **11 Science Photo Library:** Detlev Van Ravenswaay (tr). **12-13 Wikipedia:** Michael Dennig - Atlas4x4. **15 Alamy Stock Photo:** Alan Dyer / VWPics (ca). **ESO:** JAXA / Creative Commons Attribution 4.0 licence (tr). **17 Alamy Stock Photo:** Dinodia Photos (tr). **ESA:** CNES / Arianespace – Photo Optique Video du CSG – S. Martin (c). **Shutterstock.com:** oblong1 (bl). **19 Alamy Stock Photo:** MIHAI ANDRITOIU (bl); Arterra Picture Library (tr); Images of Africa Photobank (br). **Getty Images:** Jami Tarris (fbr). **20 Science Photo Library:** Dr Fred Espenak (crb, br); Babak Tafreshi (bl). **21 Brian Cassey:** (cr). **Science Photo Library:** Dr Juerg Alean (b). **22 Alamy Stock Photo:** Stockbym (r). **23 Alamy Stock Photo:** Oliver Hoffmann (l). **Science Photo Library:** Jean-Bernard Nadeau / Look At Sciences (br). **24-25 Dorling Kindersley:** Dreamstime.com: Mario Lopes / Malopes (background). **Getty Images / iStock:** ChrisGorgio (bolts). **27 Alamy Stock Photo:** George Ostertag (clb). **Getty Images / iStock:** FredericFaure (c). **Science Photo Library:** Mark Garlick (br). **30 Getty Images / iStock:** Istock (tl). **NASA:** (cl). **33 Alamy Stock Photo:** GRANGER - Historical Picture Archive (bc). **34-35 Alamy Stock Photo:** Ziga Plahutar. **37 Science Photo Library:** Shuo Wang (tr). **39 Alamy Stock Photo:** Kevin Schafer (tc). **Shutterstock.com:** MisterStock (tr). **40 Alfred Wegener Institute:** (bl). **40-41 Dorling Kindersley:** Paleogeography globes derived from original maps produced by Colorado Plateau Geosystems Inc. **42-43 Dorling Kindersley:** Dreamstime.com: Mario Lopes / Malopes (background). **Getty Images / iStock:** ChrisGorgio (bolts). **43 Shutterstock.com:** Wirestock Creators. **44 Alamy Stock Photo:** Antonio Busiello (bl). **Getty Images:** AFP PHOTO / Yasuyoshi Chiba (bc). **44-45 Dorling Kindersley:** using data from USGS / Smithsonian Institution, National Geophysical Data Center. **45 Alamy Stock Photo:** Keystone Press (bl). **Science Photo Library:** Mark Garlick (br). **46-47 Getty Images / iStock:** pxhidalgo. **47 Science Photo Library:** Jeremy Bishop (tr). **48-49 Science Photo Library:** David Weintraub. **48 Science Photo Library:** Jessica Wilson / USGS (cla). **49 Getty Images:** AFP / Lothar Slabon (crb). **naturepl.com:** Sergey Gorshkov (tl). **50 Alamy Stock Photo:** Associated Press (cl). **Getty Images:** temizyurek (cra). **Bryan Lowry / lavapix.com:** (bl). **50-51 Getty Images:** The Asahi Shimbun (bc). **51 Alamy Stock Photo:** Wei Hao Ho (cla). **Shutterstock.com:** Wirestock Creators (cra). **52 Alamy Stock Photo:** Science History Images (cra). **Shutterstock.com:** Robert Crow (c). **52-53 Shutterstock.com:** Yvonne Baur. **53 Alamy Stock Photo:** Patrick Koster (ca); Doug Perrine (fcla); tom pfeiffer (cla). **U.S. Geological Survey:** Jefffrey Judd (tr). **54-55 Bryan Lowry / lavapix.com. 56 Dorling Kindersley:** Colin Keates / Natural History Museum, London (c); Harry Taylor / Natural History Museum, London (bc). **Wikipedia:** B.Domangue (tr). **56-57 Erin Delventhal. 57 Alamy Stock Photo:** agefotostock (crb); Komkrit Suwanwela (tc); Siim Sepp (cr); Steve Pridgeon (br). **James St. John. 58-59 eyevine:** Alberto Garcia / Redux. **59 Alamy Stock Photo:** Tjetjep Rustandi (br). **Shutterstock.com:** Henk Vrieselaar (crb). **60 Alamy Stock Photo:** Laura Di Biase (tr). **60-61 BluePlanetArchive.com:** Phillip Colla. **61 Getty Images:** Arctic-Images (tc); Lepretre Pierre (tl). **62-63 Science Photo Library:** Martin Rietze. **63 Alamy Stock Photo:** Marlon Trottmann (crb). **Dorling Kindersley:** Arran Lewis (using data from NASA: Landsat 8 / School of Ocean and Earth Science and Technology Main Hawaiian Islands Multibeam Bathymetry and Backscatter Synthesis: University of Hawai'i at Ma / Hawaii Mapping Research Group) (ca). **64-65 Getty Images:** Matteo Colombo. **65 Dreamstime.com:** Barefootflyer (tr). **NASA:** ISS Crew Earth Observations experiment and the Image Science & Analysis Group, Johnson Space Center (br). **naturepl.com:** Sergio Hanquet (ca). **Shutterstock.com:** JamiesOnAMission (cr). **66 Alamy Stock Photo:** Zoonar GmbH (cra). **naturepl.com:** Guy Edwardes (tr). **66-67 Shutterstock.com:** kavram. **67 Alamy Stock Photo:** Inge Johnsson (cr); robertharding (crb). **Dorling Kindersley:** Getty Images: Kirsten Boos / EyeEm (tr). **68-69 Getty Images:** Mauro Cociglio - Turin - Italy. **70-71 Shutterstock.com:** Lynn Yeh. **71 Alamy Stock Photo:** Associated Press (br); Dinodia Photos (cr). **72-73 Alamy Stock Photo:** Associated Press. **73 Alamy Stock Photo:** Australian Associated Press (tr); GRANGER - Historical Picture Archive (crb). **USDA Forest Service (www.forestryimages.org):** (cr). **76-77 Getty Images:** AFP / JIJI PRESS. **77 Alamy Stock Photo:** Science History Images (br). **Getty Images:** AFP (tr). **NOAA:** Center for Tsunami Research (cr). **78-79 Dorling Kindersley:** Dreamstime.com: Mario Lopes / Malopes (background). **Getty Images / iStock:** ChrisGorgio (bolts). **79 Shutterstock.com:** Blue Planet Studio. **80 Alamy Stock Photo:** Chronicle (bl). **Getty Images:** Hulton Archive (br). **80-81 Getty Images / iStock:** OGphoto. **81 Alamy Stock Photo:** Grant Farquhar (fcra); imageBROKER.com GmbH & Co. KG (crb). **Dreamstime.com:** Dan Breckwoldt (cr); Nataliya Nazarova (br). **Science Photo Library:** Dirk Wiersma (tr). **83 Alamy Stock Photo:** Image Professionals GmbH (cr). **85 Dorling Kindersley:** using data from Smithsonian Institution, National Geophysical Data Center; **Shutterstock.com:** Rainer Albiez (cra); vvvita (tc); Alan Ward (crb); antony trivet photography (bc). **86-87 Getty Images / iStock:** bahadir-yeniceri. **87 Alamy Stock Photo:** Nature Picture Library (ca). **Getty Images:** Marli Miller / UCG / Universal Images Group (cr). **Getty Images / iStock:** Totajla (tr). **Shutterstock.com:** Chaikom (bc); sevenke (tl). **88 Science Photo Library:** Susumu Nishinaga (cl). **88-89 Getty Images / iStock:** jalvarezg. **89 Dorling Kindersley:** Dreamstime.com: Hai Huy Ton That (tr). **Dreamstime.com:** Joseph Gough (ca). **Shutterstock.com:** Nicolaj Larsen (cr). **SuperStock:** Egmont Strigl / Westend61 (tc). **90-91 Getty Images:** AFP / JIJI PRESS. **91 Getty Images:** Mark Gibson (cr). **Shutterstock.com:** J-B-C (tr). **92-93 Getty Images:** Roberto Moiola / Sysaworld. **93 Alamy Stock Photo:** Amos Gal (ca); Chris Mattison (cl). **NASA:** JPL-Caltech / UArizona (cr). **94 Shutterstock.com:** DCrane (clb); Jane Rix (tr). **94-95 Alamy Stock Photo:** Eva Bocek. **96 Alamy Stock Photo:** Tim Geach (cla); Piter Lenk (tc); Santi Rodriguez (c); Science History Images (cra). **97 Alamy Stock Photo:** Design Pics Inc (ca); imageBROKER.com GmbH & Co. KG (tl); GeoJuice (clb); Barry Vincent (crb). **Shutterstock.com:** Matthew Figg (bl). **98-99 naturepl.com:** Wild Wonders of Europe / Jensen. **98 Getty Images / iStock:** DurkTalsma (tr). **naturepl.com:** Franco Banfi (clb). **99 Getty Images / iStock:** S_Lew (tr). **Science Photo Library:** British Antarctic Survey (cr). **100-101 Alamy Stock Photo:** Raquel Mogado (c); Andrey Podkorytov (t). **Getty Images:** 1970s (b). **103 Science Photo Library:** Jim Steinberg (cra). **104-105 Shutterstock.com:** Blue Planet Studio. **106-107 Shutterstock.com:** Sara Winter. **106 Shutterstock.com:** 4045 (bl); Nido Huebl (cl); engel.ac (c); pcfp (bc). **108-109 Shutterstock.com:** Vadim Petrakov. **110 Alamy Stock Photo:** robertharding (bl). **Dreamstime.com:** Chase Dekker (clb). **110-111 NASA:** Joshua Stevens, using Landsat data from the U.S. Geological Survey, and soil moisture data courtesy of JPL and the SMAP Science Team (t, b). **111 Alamy Stock Photo:** US Coast Guard Photo (cr). **Getty Images:** Fatih Aktas / Anadolu Agency (tr); Ben Brewer / Bloomberg (br). **112-113 Alamy Stock Photo:** Michele Falzone. **113 Alamy Stock Photo:** ludmila yilmaz (tr). **2002 MBARI:** David Fierstein (crb). **114-115 Getty Images / iStock:** ClaudioVentrella. **116-117 Shutterstock.com:** Patrick Poendl. **117 Alamy Stock Photo:** Artenex (crb). **naturepl.com:** Alex Mustard (br). **118-119 Alamy Stock Photo:** Image Professionals GmbH. **119 Alexander Hyde:** (br). **Getty Images / iStock:** Extreme-Photographer (cr). **120 Getty Images:** Partha Pal (c). **120-121 Alamy Stock Photo:** Universal Images Group North America LLC. **121 Science Photo Library:** G. R. Roberts (crb). **122-123 Getty Images:** Andrea Comi. **122 Science Photo Library:** Martyn F. Chillmaid (tr). **123 Alamy Stock Photo:** Heritage Image Partnership Ltd (br); Westend61 GmbH (tl). **124-125 Getty Images:** Octavio Passos. **125 Alamy Stock Photo:** David Wall (cr). **Getty Images:** John Lund (tr). **Getty Images / iStock:** Thurtell (br). **126-127 Dorling Kindersley:** Dreamstime.com: Mario Lopes / Malopes (background). **Getty Images / iStock:** ChrisGorgio (bolts). **127 Alamy Stock Photo:** Mario Deambrogio. **128 Alamy Stock Photo:** agefotostock (cb). **Shutterstock.com:** Ekkaluck Sangkla (br); xpixel (ca). **128-129 Alamy Stock Photo:** Valery Voennyy (tc). **129 Alamy Stock Photo:** Arctic Images (crb); Bjrn Wylezich (bl). **Shutterstock.com:** Jiri Balek (br). **130-131 Alamy Stock Photo:** agefotostock (c). **130 Dreamstime.com:** Fokinol (cb/feldspar). **Science Photo Library:** Phil Degginger (c). **Shutterstock.com:** Moha El-Jaw (cb); J. Palys (bc). **131 Dorling Kindersley:** Gary Ombler / Oxford University Museum of Natural History (tc). **Science Photo Library:** Eye of Science (tl). **Shutterstock.com:** Ralf Lehmann (br); Vladislav S (ca). **132-133 Shutterstock.com:** Gary C. Tognoni. **133 Alamy Stock Photo:** agefotostock (ca); Ian Paterson (tc); Alan Morris (cr); Colin Harris / era-images (crb). **Shutterstock.com:** Sascha Burkard (tl). **134 Alamy Stock Photo:** Susan E. Degginger (tr). **Dorling Kindersley:** Colin Keates / Natural History Museum, London (tl). **Dreamstime.com:** Viktor Nikitin (br); Pancaketom (cla). **134-135 Dreamstime.com:** Kostrykina Myroslava (c). **135 Alamy Stock Photo:** Panther Media GmbH (bc). **Dreamstime.com:** Natalis66 (bl). **Science Photo Library:** Dirk Wiersma (crb). **Shutterstock.com:** Karel Funda (br). **136 Alamy Stock Photo:** The Natural History Museum (tl). **Science Photo Library:** Ashley Cooper (cla); Kaj R. Svensson (clb); Jean-Claude Revy, ISM (bl). **136-137 Shutterstock.com:** Jeroen Mikkers. **137 Science Photo Library:** Steve Lowry (tr). **138-139 Alamy Stock Photo:** YAY Media AS. **139 Alamy Stock Photo:** imageBROKER.com GmbH & Co. KG (br); mauritius images GmbH (cr). **Science Photo Library:** Eye of Science (tl); Wim van Egmond (clb); Nature Picture Library (cra). **Shutterstock.com:** hanif66 (cla). **140-141 Alamy Stock Photo:** Mario Deambrogio. **140 Science Photo Library:** Microckscopica (clb). **141 Alamy Stock Photo:** Igor Petrushenko (tc). **Dorling Kindersley:** Dreamstime.com: Vlad3563 (tr). **Dreamstime.com:** Chormail (cr). **Getty Images:** John W Banagan (br). **142-143 Alamy Stock Photo:** Phil Degginger (c). **142 Alamy Stock Photo:** Bjrn Wylezich (bc). **Dorling Kindersley:** Ruth Jenkinson / RGB Research Limited (bl); Colin Keates / Natural History Museum, London (br). **143 Dorling Kindersley:** Dreamstime.com: Christophe Testi (tr); Fotolia: apttone (ca). **Science Photo Library:** Javier Trueba / MSF (cr). **Shutterstock.com:** Yeti studio (cra). **144 Alamy Stock Photo:** Wirestock, Inc. (tr). **Shutterstock.com:** Viktoria Prusakova (cl). **144-145 Shutterstock.com:** Dan Olsen. **145 Dorling Kindersley:** Colin Keates / Natural History Museum, London (tc). **Science Photo Library:** Mark A. Schneider (cr); Dirk Wiersma (br). **147 Alamy Stock Photo:** PjrRocks (cr); Bill Truran (crb). **James Van Fleet:** (cra). **148 Dorling Kindersley:** Dreamstime.com: Fokinol (cla); Colin Keates / Natural History Museum, London (fcl)(ca)(cr)(fbl)(cb)(bc). **148-149 Dorling Kindersley:** iStock: rusm (background). **149 Dorling Kindersley:** Colin Keates / Natural History Museum, London (tl)(cla)(cl)(tc)(ftr). **150-151 Getty Images:** SunChan (cb). **150 Alamy Stock Photo:** Phil Degginger (cl). **Dorling Kindersley:** Colin Keates / Natural History Museum (ca); Tim Parmenter / Natural History Museum, London (tc). **Shutterstock.com:** STUDIO492 (cla). **151 Alamy Stock Photo:** Phil Degginger (cla); Gemma Fletcher (tl). **Dorling Kindersley:** Dreamstime.com: Bjrn Wylezich (cra); Tim Parmenter / Natural History Museum, London (tc); Richard Leeney / Holts Gems, Hatton Garden (c); Colin Keates / Natural History Museum, London (crb)(bl)(bc). **Shutterstock.com:** BGStock72 (cl); horiyan (tr); J. Palys (cb). **152 Shutterstock.com:** Bjoern Wylezich (bl). **152-153 De Beers:** (c). **153 Alamy Stock Photo:** Skipping Cricket (br); David Tadevosian (tc). **154-155 Shutterstock.com:** Albert Russ. **155 Shutterstock.com:** DedMityay (br); Bjoern Wylezich (tr); Bjoern Wylezich (cra); Sebastian Janicki (cr); SHTRAUS DMYTRO (crb). **156-157 Shutterstock.com:** lkpro. **158 Science Photo Library:** (cb); Marek Mis (cra); STEVE GSCHMEISSNER (ca)(cr)(cl)(clb). **Shutterstock.com:** Nicola Pulham (bl); Natalia van D (tr). **158-159 Alamy Stock Photo:** jo ingate (bc). **159 Science Photo Library:** INNERSPACE IMAGING (tr); UCL, GRANT MUSEUM OF ZOOLOGY (cla); SUSUMU NISHINAGA (cr)(br). **Shutterstock.com:** New Africa (ca). **160-161 Dorling Kindersley:** Dreamstime.com: Mario Lopes / Malopes (background). **Getty Images / iStock:** ChrisGorgio (bolts). **161 Kenneth G. Libbrecht. 162-163 Dreamstime.com:** Studio023. **163 Science Photo Library:** NASA'S GODDARD SPACE FLIGHT CENTER (tl). **Shutterstock.com:** PhotoVisions (ca). **165 Dorling Kindersley:** Dreamstime.com: Elena Gurdina (cr). **Science Photo Library:** Steve Gschmeissner (crb). **Shutterstock.com:** Kay Cee Lens and Footages (cra). **166-167 Dreamstime.com:** Feelgoodsk. **167 Alamy Stock Photo:** Rapt.Tv (tr). **168-169 Getty Images:** Greg Pease. **169 Alamy Stock Photo:** Science Photo Library (cra). **Dorling Kindersley:** Dreamstime.com: View7 (cla). **170-171 Dorling Kindersley:** using data from www.worldclim.org. **171 Alamy Stock Photo:** hanohikirf (cb). **Getty Images:** Emad aljumah (ca). **Getty Images / iStock:** Vagabondering Andy - Andy Doyle (bc). **Shutterstock.com:** Jane Rix (tc). **172-173 Dorling Kindersley:** using data from www.worldclim.org. **173 Alamy Stock Photo:** Colin Harris / era-images (bl). **naturepl.com:** Christophe Courteau (ca). **NOAA:** GOES Project Science Office (crb). **Shutterstock.com:** Deliris (tc); Ondrej Prosicky (cl); Bassel Rachid (cr). **174-175 Science Photo Library:** Karsten Schneider. **175 Science Photo Library:** NOAA (bc). **176 Alamy Stock Photo:** NG Images (tr). **176-177 Alamy Stock Photo:** Timothy Smith. **177 Alamy Stock Photo:** Robert Adrian Hillman (tl). **178 ESA:** NASA–A. Gerst (cb). **178-179 Shutterstock.com:** elRoce. **179 Alamy Stock Photo:** US Air Force Photo (tr). **Getty Images / iStock:** adrianorgza (tc). **Getty Images:** Warren Faidley (tl). **NOAA:** Digital GeoZone (cra). **180-181 Brian A. Morganti / StormEffects.com. 181 NASA:** Earth Observatory images by Jesse Allen, using Landsat data from the U.S. Geological Survey (crb). **183 Dreamstime.com:** Richair (tc). **184 Shutterstock.com:** Derek Beattie Images (tr); Lukas Jonaitis (l); kristof lauwers (cr). **184-185 Shutterstock.com:** Frannyanne (c); Brian A Jackson (b); Kay Cee Lens and Footages. **185 Alamy Stock Photo:** Delphotos (br). **Science Photo Library:** Mike Hollingshead (cra). **Shutterstock.com:** alybaba (clb); paul prescott (bc); Sebastian Hulse (tl); Phil Semmens (tr); kostin77 (cr). **186-187 Science Photo Library:** Roger Hill. **188 Dreamstime.com:** Maciej Czekajewski (bl); Natalia Pushchina (clb). **188-189 Shutterstock.com:** HoleInTheBox. **189 Alamy Stock Photo:** Dave Bevan (tl); Jon G. Fuller / VWPics (tr). **Getty Images:** JC Patricio (tc). **191 Alamy Stock Photo:** James Lewis / Stockimo (br). **Dorling Kindersley:** Dreamstime.com: James Wheeler / Souvenirpixels (cr). **192-193 Shutterstock.com:** Naufal MQ. **193 Shutterstock.com:** Tomasz Duma (tr); SakSa (tl); Dean Pennala (ca). **194 Dan Robinson:** (r). **195 Alamy Stock Photo:** Nature Picture Library (cr). **Dreamstime.com:** Baloncici (br). **Dan Robinson. 196 Kenneth G. Libbrecht. 197 Kenneth G. Libbrecht. Science Photo Library:** Kenneth Libbrecht (cra). **198 Alamy Stock Photo:** Richard Lewis (tc); Antonella Lussardi (bl). **Science Photo Library:** Nature's Faces / Science Science (cla); Pekka Pariainen (clb). **198-199 Alamy Stock Photo:** David Forster. **199 Alamy Stock Photo:** Mark Pink (tr). **200-201 Getty Images:** Allan Davey. **200 Alamy Stock Photo:** Thierry Grun (cla); John Sirlin (clb). **Getty Images:** Mike Lyvers (bl). **NOAA:** GOES East (tl). **202-203 Marko Koroec. 204-205 Alamy Stock Photo:** John Sirlin. **205 Getty Images / iStock:** Dr_Microbe (cra). **Science Photo Library:** Digital Globe (cla); NOAA (ca). **206 Greenpeace:** Gesellschaft fr kologische Forschung (ca). **207 Alamy Stock Photo:** Science History Images (bc). **Greenpeace:** Gesellschaft fr kologische Forschung (ca). **Shutterstock.com:** Vladimir Endovitskiy (br). **208-209 Dorling Kindersley:** Dreamstime.com: Mario Lopes / Malopes (background). **Getty Images / iStock:** ChrisGorgio (bolts). **209 Getty Images:** imageBROKER / Peter Giovannini (c). **210 Alamy Stock Photo:** RooM the Agency (cb). **Wikipedia:** MARUM- Center for Marine Environmental Sciences, University of Bremen (c). **211 Alamy Stock Photo:** Branko Devic (br). **Dorling Kindersley:** Dreamstime.com: Solarseven (cb). **Getty Images:** Roger Ressmeyer / Corbis / VCG (cra). **212 Alamy Stock Photo:** blickwinkel (cra). **212-213 naturepl.com:** Gary Bell / Oceanwide / Minden. **213 Getty Images / iStock:** Dmitriy Sidor (cla). **Science Photo Library:** Dirk Wiersma (ca). **214 Getty Images / iStock:** CribbVisuals (bl); wanderluster (cl). **216 Alamy Stock Photo:** Nick Upton (tc). **Getty Images / iStock:** Xurzon (clb). **Shutterstock.com:** Aureliy (bl); Lucas Leuzinger (br). **217 Getty Images / iStock:** Frank DeBonis (bc); Paolo Graziosi (tl); Shunyu Fan (tc); DieterMeyrl (cra); pierivb (crb); dennisvdw (bl). **218-219 Shutterstock.com:** Linda Szeto. **218 Getty Images / iStock:** Oleh_Slobodeniuk (br). **Shutterstock.com:** Mats Brynolf (bl); Nata Naumovec (cr). **219 Getty Images:** Its About Light / Design Pics (br). **naturepl.com:** Suzi Eszterhas / Minden (c). **Shutterstock.com:** Jukka Jantunen (bl). **220-221 Getty Images / iStock:** Schroptschop. **220 Getty Images / iStock:** Avalon_Studio (c). **Shutterstock.com:** Jack Bell Photography (bl); Kevin Wells Photography (br). **221 Getty Images / iStock:** JasonOndreicka (bl); standret (cl); pchoui (cr). **Shutterstock.com:** Erni (br). **222-223 Getty Images / iStock:** Matt_Gibson. **222 Getty Images / iStock:** Elizabeth Nunn (bl); Wirestock (cr). **Shutterstock.com:** WildMedia (br). **223 Alamy Stock Photo:** Duncan Usher (cl). **Getty Images / iStock:** Kaphoto (br); Michael_Conrad (bl). **224-225 Getty Images:** Paul Biris. **224 Guillermo Ferraris and Mariella Superina:** (bl). **Getty Images / iStock:** mauinow1 (br). **225 Alamy Stock Photo:** blickwinkel (cr); Martin Harvey (cl). **Getty Images:** S. Daniel McPhail / 500px (br). **Getty Images / iStock:** NormaZaro (c). **Shutterstock.com:** Anan Kaewkhammul (bl). **226-227 Getty Images:** Paul Biris. **226 Dorling Kindersley:** Dreamstime.com: Eastmanphoto (br). **Getty Images:** Ignacio Palacios (crb). **Shutterstock.com:** Medolka (cra); Matt Starling Photography (bl, bc). **227 Dorling Kindersley:** Andrew Beckett (Illustration Ltd) (fbr). **Getty Images:** cinoby (crb). **naturepl.com:** Daniel Heuclin (ca). **Science Photo Library:** Vincent Amouroux, Mona Lisa Production (bl). **Shutterstock.com:** MedMounirPic (cla). **228-229 Shutterstock.com:** Yusnizam Yusof. **228 Getty Images / iStock:** agustavop (cr); Lillian Tveit (bl). **Shutterstock.com:** Kurit afshen (br). **229 Alamy Stock Photo:** Marc Anderson (c); Natalia Golovina (bl). **Shutterstock.com:** Usanee (br). **230-231 Getty Images:** Christophe Paquignon. **230 Getty Images:** Bryngelzon (bl). **Shutterstock.com:** Stu Porter (tr). **231 Alamy Stock Photo:** Imagebroker (cla); robertharding (bc). **Getty Images:** Nick Dale / 500px (ca). **Shutterstock.com:** Zeinab Alameh (cra). **232-233 Alamy Stock Photo:** Markus Thomenius. **232 Shutterstock.com:** Fabian Ponce Garcia (bl); Martin Mecnarowski (tl); Jana Troupova (bc); KLiK Photography (br). **233 Alamy Stock Photo:** Arterra Picture Library (clb); Minden Pictures (crb). **Shutterstock.com:** Carlos Sala Fotografia (ca); Milton Rodriguez (bc). **234-235 naturepl.com:** Rhonda Klevansky. **234 Alamy Stock Photo:** David G Richardson (cr). **naturepl.com:** David Fleetham (br). **Shutterstock.com:** David Steele (bl). **235 Alamy Stock Photo:** Bob Gibbons (cra). **Dreamstime.com:** Gaspoll (crb). **naturepl.com:** Christophe Courteau (cb). **Andrea Nixon / AndyNixPix:** (cla). **Steve Woodhall / butterflygear.co.za:** (bl). **236-237 Alamy Stock Photo:** imageBROKER.com GmbH & Co. KG. **236 naturepl.com:** Nick Garbutt (br). **Shutterstock.com:** buteo (cb); Brian Lasenby (clb). **237 Alamy Stock Photo:** Christian Dietz (bl); Panoramic Images (cb). **Getty Images:** imageBROKER / Peter Giovannini (cla). **Getty Images / iStock:** Alex Potemkin (tr); Mary Swift (tl). **Shutterstock.com:** miroslav chytil (crb); Wiro.Klyngz (cra). **238-239 BluePlanetArchive.com:** Reinhard Dirscherl. **239 naturepl.com:** Doug Allan (cr); Brandon Cole (tr); David Shale (crb). **240-241 Getty Images / iStock:** undefined undefined. **242-243 Dorling Kindersley:** Dreamstime.com: Mario Lopes / Malopes (background). **Getty Images / iStock:** ChrisGorgio (bolts). **246 Dorling Kindersley:** iStock: Ron and Patty Thomas (crb). **Shutterstock.com:** Lukas Bischoff Photograph (cra). **246-247 Dreamstime.com:** Tsvibrav (tc); Wirestock (c). **Getty Images:** Ignacio Palacios (bc). **247 Dreamstime.com:** Barelkodotcom (tc); Vladimir Rodin (cb). **Getty Images:** hadynyah (cl). **Shutterstock.com:** Misne (ca); Vaclav P3k (cra). **248 Shutterstock.com:** lavizzara (cr); On_the_road (tr); zevana (clb). **248-249 Dreamstime.com:** Mariusz Prusaczyk (bc). **249 Alamy Stock Photo:** robertharding (crb); Westend61 GmbH (bc). **Getty Images / iStock:** andryslukowski (cla); bjdlzx (cra). **Shutterstock.com:** Brastock (clb); Valerii_M (tl). **250 Dreamstime.com:** Joshua Wanyama (br). **Science Photo Library:** Planetobserver (tl). **Shutterstock.com:** Igor Kovalenko (cra). **250-251 Getty Images / iStock:** jono0001 (c). **Shutterstock.com:** Katvic (tc). **251 Alamy Stock Photo:** Universal Images Group North America LLC (br). **Science Photo Library:** Planetobserver (tr)(bl). **Shutterstock.com:** Federica Cordero (cr); Max Forgues (ca). **252 Alamy Stock Photo:** John Martin - Fotografo (bl); Westend61 GmbH (tr). **Shutterstock.com:** David A Knight (br). **253 Alamy Stock Photo:** Danita Delimont (bl); Chris Howes / Wild Places Photography (tr); Nature Picture Library (cr). **Dave Bunnell:** (br). **Getty Images / iStock:** Eder Maioli (tc). **Getty Images:** Bernard Friel / UIG (cla). **Shutterstock.com:** Nora Yusuf (cl). **254 Getty Images:** Feng Wei Photography (cb); Whitworth Images (bl). **Getty Images / iStock:** Laszlo Peto (tr); rchphoto (br). **254-255 Getty Images / iStock:** Patrick Jennings (c). **255 Getty Images / iStock:** Boogich (br). **Shutterstock.com:** Belikart (cr); stacyarturogi (tl); Michael Mantke (tr); Andrea Izzotti (bl). **256 Alamy Stock Photo:** Dan Leeth (crb). **Getty Images:** Mark Newman (tr). **Shutterstock.com:** Wirestock Creators (cla). **256-257 Getty Images:** Alexander Piragis (ca). **Science Photo Library:** Prof. Stewart Lowther (bc). **257 Alamy Stock Photo:** Media Drum World (cl); Moodboard Stock Photography (tr); tom pfeiffer (br). **Getty Images:** Gianni Sarasso (cr). **Science Photo Library:** Martin Rietze (tc). **258 Alamy Stock Photo:** NASA Photo (cla). **NASA:** GSFC / OIB (tr). **Shutterstock.com:** Eugene Ga (crb). **258-259 123RF.com:** muraviov (bc). **Alamy Stock Photo:** Maridav (ca). **259 Alamy Stock Photo:** Nacho Calonge (cl); Universal Images Group North America LLC (cr). **Getty Images:** Raimund Linke (br); MisoKnitl (tc). **Shutterstock.com:** ole (tr). **260 Dorling Kindersley:** Harry Taylor / Natural History Museum, London. **Shutterstock.com:** www.sandatlas.org (tc); Yes058 Montree Nanta (cra). **261 James St. John. 263 Dorling Kindersley:** Colin Keates / Natural History Museum, London (cb/bryozoan); Harry Taylor / Natural History Museum, London (cr)(cb)(bl). **Shutterstock.com:** Yes058 Montree Nanta (ca); Yes058 Montree Nanta (bc). **264 Dorling Kindersley:** Colin Keates / Natural History Museum, London (crb). **Shutterstock.com:** Cagla Acikgoz (cla). **265 Dorling Kindersley:** Colin Keates / Natural History Museum, London (c). **268 Alamy Stock Photo:** The Natural History Museum (cl). **Dorling Kindersley:** 123RF: Micha Baraski / 123rf (tr); Colin Keates / Natural History Museum, London (bc). **Getty Images / iStock:** Minakryn Ruslan (crb). **Science Photo Library:** Science Stock Photography (ca). **Shutterstock.com:** Jack Dagley Photography (fcl); Reload Design (fcr). **268-269 Dorling Kindersley:** Tim Parmenter / Natural History Museum, London (gems). **269 Alamy Stock Photo:** Halyna Kubei (cb); slaughteredlamb (ca); SBS Eclectic Images (clb). **Dorling Kindersley:** Ruth Jenkinson / Holts Gems (tr/Kunzite); Colin Keates / Natural History Museum, London (cl/jadeite)(tc). **Dreamstime.com:** Madalin Stancu (cr). **Getty Images / iStock:** VvoeVale (cl). **Shutterstock.com:** fotoscool (crb). **270 Dorling Kindersley:** Colin Keates / Natural History Museum, London (tr); Gary Ombler / Senckenberg Gesellschaft Fuer Naturforschung Museum (tc); Colin Keates / Natural History Museum (c); Gary Ombler / Swedish Museum of Natural History (clb). **Dr David J. Ward:** (cla). **271 Dorling Kindersley:** Andy Crawford / Senckenberg Nature Museum (tl); Trustees of the Natural History Museum, London (ftl); Gary Ombler / Senckenberg Gesellschaft Fuer Naturforschung Museum (cl); Harry Taylor / Natural History Museum, London (cl/skull); Gary Ombler / Oxford Museum of Natural History (tc); Dreamstime.com: Ken Backer (ca); Colin Keates / Natural History Museum, London (ca/dodo)(ca/egg)(br)(tr/fern)(cra)(cr)(crb/leaf)(fbr); Gary Ombler / Oxford Museum of Natural History (c); Harry Taylor / Hunterian Museum University of Glasgow (bc); Dreamstime.com: Bjrn Wylezich (crb/amber). **Dr David J. Ward. 272-273 Dorling Kindersley:** using data from NASA and NOAA.